Climate Change and Anthropogenic Impacts on Health in Tropical and Subtropical Regions

Ahmed Karmaoui
Southern Center for Culture and Sciences, Morocco

Abdelkrim Ben Salem
Southern Center for Culture and Sciences, Morocco & Research Center of Plant and Microbial Biotechnologies, Biodiversity, and Environment (BioBio), Faculty of Sciences, Mohammed V University, Rabat, Morocco

Ashfaq Ahmad Shah
National University of Sciences and Technology, Pakistan

A volume in the Advances in Environmental Engineering and Green Technologies (AEEGT) Book Series

IGI Global
DISSEMINATOR OF KNOWLEDGE

Published in the United States of America by
IGI Global
Engineering Science Reference (an imprint of IGI Global)
701 E. Chocolate Avenue
Hershey PA, USA 17033
Tel: 717-533-8845
Fax: 717-533-8661
E-mail: cust@igi-global.com
Web site: http://www.igi-global.com

Library of Congress Cataloging-in-Publication Data

Names: Karmaoui, Ahmed, 1982- editor. | Ben Salem, Abdelkrim, 1981- editor. | Shah, Ashfaq Ahmed, 1986- editor.
Title: Climate change and anthropogenic impacts on health in tropical and subtropical regions / Ahmed Karmaoui, Abdelkrim Ben Salem, and Ashfaq Ahmed Shah, editors.
Description: Hershey, PA : Engineering Science Reference, [2020] | Includes bibliographical references and index. | Summary: "This book provides vital research on the association of environmental pollutants and global warming with viruses in tropical regions. It also explores the developments and trends in these areas of medicine and ecology, as well as prevention strategies to be used for educational and sensitization purposes"-- Provided by publisher.
Identifiers: LCCN 2019037697 (print) | LCCN 2019037698 (ebook) | ISBN 9781799821977 (hardcover) | ISBN 9781799821984 (paperback) | ISBN 9781799821991 (ebook)
Subjects: LCSH: Tropical medicine. | Medical climatology. | Climatic changes--Health aspects. | Environmental health. | Human beings--Effect of climate on.
Classification: LCC RC961.5 .C59 2020 (print) | LCC RC961.5 (ebook) | DDC 614.2/23--dc23
LC record available at https://lccn.loc.gov/2019037697
LC ebook record available at https://lccn.loc.gov/2019037698

This book is published in the IGI Global book series Advances in Environmental Engineering and Green Technologies (AEEGT) (ISSN: 2326-9162; eISSN: 2326-9170)

British Cataloguing in Publication Data
A Cataloguing in Publication record for this book is available from the British Library.

All work contributed to this book is new, previously-unpublished material.
The views expressed in this book are those of the authors, but not necessarily of the publisher.

For electronic access to this publication, please contact: eresources@igi-global.com.

Advances in Environmental Engineering and Green Technologies (AEEGT) Book Series

ISSN:2326-9162
EISSN:2326-9170

MISSION

Growing awareness and an increased focus on environmental issues such as climate change, energy use, and loss of non-renewable resources have brought about a greater need for research that provides potential solutions to these problems. Research in environmental science and engineering continues to play a vital role in uncovering new opportunities for a "green" future.

The **Advances in Environmental Engineering and Green Technologies (AEEGT)** book series is a mouthpiece for research in all aspects of environmental science, earth science, and green initiatives. This series supports the ongoing research in this field through publishing books that discuss topics within environmental engineering or that deal with the interdisciplinary field of green technologies.

COVERAGE

- Pollution Management
- Electric Vehicles
- Industrial Waste Management and Minimization
- Cleantech
- Radioactive Waste Treatment
- Waste Management
- Sustainable Communities
- Water Supply and Treatment
- Renewable Energy
- Air Quality

IGI Global is currently accepting manuscripts for publication within this series. To submit a proposal for a volume in this series, please contact our Acquisition Editors at Acquisitions@igi-global.com or visit: http://www.igi-global.com/publish/.

The Advances in Environmental Engineering and Green Technologies (AEEGT) Book Series (ISSN 2326-9162) is published by IGI Global, 701 E. Chocolate Avenue, Hershey, PA 17033-1240, USA, www.igi-global.com. This series is composed of titles available for purchase individually; each title is edited to be contextually exclusive from any other title within the series. For pricing and ordering information please visit http://www.igi-global.com/book-series/advances-environmental-engineering-green-technologies/73679. Postmaster: Send all address changes to above address.

Titles in this Series

For a list of additional titles in this series, please visit:
https://www.igi-global.com/book-series/advances-environmental-engineering-green-technologies/73679

Physiological Aspects of Imprinting and Homing Migration in Salmon Emerging Research and Opportunities
Hiroshi Ueda (Hokkaido University, Japan)
Engineering Science Reference • © 2020 • 200pp • H/C (ISBN: 9781799820543) • US $175.00

Artificial Intelligence and IoT-Based Technologies for Sustainable Farming and Smart Agriculture
Pradeep Tomar (Gautam Buddha University, India) and Gurjit Kaur (Delhi Technological University, India)
Engineering Science Reference • © 2020 • 350pp • H/C (ISBN: 9781799817222) • US $215.00

Predicting, Monitoring, and Assessing Forest Fire Dangers and Risks
Nikolay Viktorovich Baranovskiy (National Research Tomsk Polytechnic University, Russia)
Engineering Science Reference • © 2020 • 417pp • H/C (ISBN: 9781799818670) • US $235.00

Effects of Emerging Chemical Contaminants on Water Resources and Environmental Health
Victor Shikuku (Kaimosi Friends University College, Kenya)
Engineering Science Reference • © 2020 • 350pp • H/C (ISBN: 9781799818717) • US $225.00

Waste Management Techniques for Improved Environmental and Public Health Emerging Research and Opportunities
Sang-Bing Tsai (Zhongshan Institute, University of Electronic Science and Technology of China, China & Wuyi University, China) Zhengxi Yuan (Zhongshan Institute, University of Electronic Science and Technology of China, China) Jian Yu (Research Center for Environment

For an entire list of titles in this series, please visit:
https://www.igi-global.com/book-series/advances-environmental-engineering-green-technologies/73679

IGI Global
DISSEMINATOR OF KNOWLEDGE

701 East Chocolate Avenue, Hershey, PA 17033, USA
Tel: 717-533-8845 x100 • Fax: 717-533-8661
E-Mail: cust@igi-global.com • www.igi-global.com

Table of Contents

Detailed Table of Contents

Conferring to the Global Risk Index, Pakistan is ranked as the 7th most susceptible country to the inexorable influence of climate change. Before this century ends, the annual mean temperature in Pakistan is expected to rise from 3°C to 5°C for a focal worldwide discharge situation. Usually, annual precipitation is not relied upon to have a critical long haul pattern. Ocean level is relied upon to ascend further by 60 centimeters. All these climatic events are likely to disrupt the economy, lives, and the socio-political aspects of human life. Pakistan has already witnessed massive loss in terms of human, infrastructural, and economic aspects. The chapter is designed to understand both the direct and indirect health risks associated with frequent climatic events like floods, drought, and heat waves in Pakistan. After analyzing the available literature, it was observed that floods and drought have direct and indirect health risks associated with them while in case of heat waves, health risks cannot be established precisely as multiple variables are involved, playing a significant role.

The progress on reducing stunting is rather slow in Pakistan despite significant reductions in poverty which begs a question why Pakistan has been unable to make significant strides in improving nutrition indicators for children over the past few decades. Despite the recognized importance of the problem in national and international forums, research on determinants of child stunting in Pakistan is scarce, especially in the context of the role of mother's health, education, and empowerment in determining a child's nutrition status. Therefore, this chapter incorporates the mother's health, education, and empowerment-related factors in determining the factors that affect child stunting in Pakistan. Using simple OLS methodology on DHS (2012-13) dataset for Pakistan, the authors' results show that improvement in mother's health, women empowerment, and women's education are likely to reduce stunting. Moreover, better hygiene and better food intake also reduce stunting among children in Pakistan.

Home Garden is a complex sustainable land use system that combines multiple farming components of the homestead and provides environmental services, household needs, and employment and income generation opportunities to the households. Predicted climate changes have serious implications for crop and livestock yields particularly in tropical regions. Home garden may act as a cushion to the adverse climate shocks. There is dearth of in-depth study of home garden ecosystem in India. Authors selected 100 households in Garhbeta-1 block, which is in the dry zone in the district of Paschim Medinipur in West Bengal for this study. The main objectives of this chapter include: (a) identifying key characteristics of the home garden, (b) identifying the pattern of climate change from the household perceptions and the changes made in the home gardens. (c) identifying different diseases of plants and animals in home gardens in the study area (d) managing diseases of plants in home gardens, (e) identifying different problems/ constraints in home.

Section 2
Climate Change and Neglected Tropical Diseases Associations

This chapter has proposed a systematic method to design mathematical models. These models have been associated with counting of white blood cells, counting of red blood cells, population of mosquitoes, and counting of foreign bodies like virus, bacteria, and parasite in a human body. Interpretations for critical points or equilibrium points have been given for network systems of differential equations associated with models. A practical method of applying these interpretations in administrating medicines to get control over diseases has been given. Order of priority in three types of critical points, namely, stable critical points, unstable critical points, and asymptotically stable critical points, has been given. Conversions of differential equations of models into integral equations and applying Picard's iteration method to solve integral equations have been explained. A step-by-step approach has been used in designing models, solving models, and interpreting solutions of models for tropical diseases.

Section 3
Climate Change and Anthropogenic Impacts on Some Tropical Neglected Diseases From Two Vulnerable Continents: Africa and Asia

Strongyloidiasis is a human parasitic ailment brought about by a whiplike nematode worm called Strongyloides stercoralis. Most humans get the infection by coming in contact with contaminated soil whereby the tiny worms penetrate the skin and enter the bloodstream from where it passes through the right side of the heart and lungs to mouth, stomach, and small intestines. Replication inside tainted people enables the disease to persevere for quite a long time. Strongyloides can cause a hyperinfection disorder which causes side effects in numerous organ frameworks, including the central nervous system that can prompt death if untreated. The diagnosis is made by blood and stool tests. This chapter is intended to draw a more precise picture of the global prevalence, diagnosis, and risk factors for S. stercoralis. The chapter also discusses the diagnostic approaches for detecting the infection, the morbidity caused and the recommended management. It further discusses some of the reasons why this infection is so neglected and the consequence of this for the estimated global prevalence.

In 2009, the Moroccan Ministry of Health in collaboration with the World Health Organization confirmed that the transmission of schistosomiasis had ceased in all previously endemic provinces of Morocco. This achievement couldn't have been achieved without the effective participation and engagement of local communities. In this chapter, studies of community participation strategies for schistosomiasis control in Akka oasis were reviewed. Southern Morocco was known to be a foci site for urinary schistosomiasis since 1937. Combined efforts between health officials and local community participation have achieved the elimination of this neglected tropical disease, where the last cases in these foci were recorded in the early 2000s. The results of this chapter can be used to increase awareness of the need for community participation in controlling neglected tropical diseases such as schistosomiasis.

The current chapter deals with one of the most neglected tropical diseases in Morocco, the cutaneous leishmaniasis. It is based on 10-year research (2010-2017) on the evolution of leishmaniasis taking climate change into account. Epidemiological and climatological data were collected from different administrations. The Geographic Information System (GIS) is chosen for interpolation, space-time analysis of climate data and map creation. The SPSS software was used for statistical analysis and to establish the relationship between Leishmaniasis and climatic conditions. Results show that the maximum number of cases is recorded in 2010 with 4,407 people affected while the low number is recorded in 2014 with 18 cases. Results also show a clear link between climatic factors and the incidence of the disease. The distribution of the disease in the province is influenced by maximum temperature, aridity, and vegetation cover. Additionally, anthropogenic factors play a significant role in explaining the emergence or re-emergence of leishmaniasis in the region.

Preface

Many studies around the world have proven that industrial activities leave large quantities of greenhouse gases, mainly carbon dioxide (CO2), causing climate change. This change has led to tremendous and varied effects on the health of living things and on humans. The rise in temperature and sea level, low precipitation rates and an increase in the frequency of floods caused a number of diseases. Anthropogenic factors like urbanization, deforestation, and pollution are among the factors that contribute to exacerbating epidemic diseases. Changes in the distribution of the neglected tropical diseases (NTDs) including vector-borne diseases due to air, soil and water pollution are expected. The prevalence of these diseases has become significant, especially in areas where there is social and economic vulnerability. This is due to inadequate health systems, lack of resources, and malnutrition.

In this book, the relationship between climate and health is stated with a focus on some neglected tropical diseases. It can equip new generations, practitioners, populations, and research projects in the field. Mainly, four types of diseases Leishmaniasis, Schistosomiasis, Strongyloidiasis, and Malaria were explored. These include epidemiology, climatology, and ecology, and medicine. The NTDs in areas of low-income and middle-income are caused by bacteria, viruses, protozoa, or parasitic worms (Helminthes). The poor living conditions contribute to expanding the infections. As their name indicates, these diseases are neglected which affects its eradication. Some of these diseases can be treated and others (Visceral Leishmaniasis) can cause death.

The impact factors are of climatic and anthropogenic origin. The role of climate in its influence on vectors and reservoirs of tropical diseases cannot be denied. However, there are human factors that are certainly to have an impact on the spread of these diseases. In fact, a number of human activities represented in irrigation networks in the agricultural field and sanitation in cities, especially villages, have an impact on these diseases. Also, the urban expansion and the spread of random housing that creates solid waste accumulation are among the factors that change the environment of pathogen vectors and reservoirs, contributing to the spread of diseases. In addition, migration and means of transportation are among the reasons for this spread of NTDs.

These diseases affect more than a billion people around the world and contribute to deepening the poverty crisis and workday losses for thousands of workers and students. The tropical regions, especially those with a high prevalence of tropical diseases, are making great efforts to control them. This is through advancing the health and environmental situation. Tropical diseases are considered complex health problems due to the variety of causes and factors leading to them. To understand the relationship of climate change, anthropogenic activities, and some neglected tropical diseases, further research and fieldwork must be increased in different disciplines of epidemiology, climatology, environment, and medicine. Also, it is necessary to create and develop models that explain the spread of the diseases, monitor and anticipate them using climate variables. The book is organized in seven chapters.

Chapter 1 explores the climate change and health impacts in Pakistan. Conferring to the Global Risk Index, Pakistan is ranked as the 7th most susceptible country to the inexorable influence of climate change. Before this century ends, the annual mean temperature in Pakistan is expected to rise from 3°C to 5°C for a focal worldwide discharge situation. Usually, annual precipitation is not relied upon to have a critical long haul pattern. The ocean level is relied upon to ascend further by 60 centimeters. All these climatic events are likely to disrupt the economy, lives and the socio-political aspects of human life. Pakistan has already witnessed massive loss in terms of human, infrastructural and economic aspects. The chapter is designed to understand both the direct and indirect health risks associated with frequent climatic events like floods, drought and heatwaves in Pakistan. After analyzing the available literature, it was observed that floods and drought have direct and indirect health risks associated with them while in the case of heatwaves, health risks cannot be established precisely as multiple variables are involved, playing a significant role.

Chapter 2 highlights maternal determinants of childhood stunting: the case of Pakistan. The progress on reducing stunting is rather slow in Pakistan despite significant reductions in poverty which begs a question why Pakistan has been unable to make significant strides in improving nutrition indicators for children over the past few decades. Despite the recognized importance of the problem on national as well as international forums, research on determinants of child stunting in Pakistan is scarce, especially in the context of the role of mother's health, education, and empowerment in determining a child's nutrition status. Therefore, the objective of this study is to incorporate the mother's health, education and empowerment-related factors in determining the factors that affect child stunting in Pakistan. Using simple OLS methodology on DHS (2012-13) dataset for Pakistan, the results show that improvement in mother's health, women empowerment, and women's education are likely to reduce stunting. Moreover, better hygiene and better food intake also reduce stunting among children in Pakistan.

Chapter 3 studied the climate change and diseases of plants and animals in home gardens of west Bengal, India. Home Garden" is a complex sustainable land-use system that combines multiple farming components of the homestead and provides environmental services, household needs, and employment and income generation opportunities to the households. Predicted climate changes have serious implications for crop and livestock yields particularly in tropical regions. The home garden may act as a cushion to the adverse climate shocks. There is a dearth of an in-depth study of the home garden ecosystem in India. We have selected 100 households in Garhbeta-1 block, which is in the dry zone in the district of Paschim Medinipur in West Bengal for our study. The main objectives of the paper include (a) identification of the key characteristics of the home garden, (b) to identify the pattern of climate change from the household perceptions and the changes made in the home gardens. (c) To identify different diseases of plants and animals in home gardens in the study area (d) Management of diseases of plants in home gardens, (e) To identify different problems/ constraints in a home.

Chapter 4 investigates the network modeling on tropical diseases vs. climate change. This chapter has proposed a systematic method to design mathematical models. These models have been associated with counting of white blood cells, counting of red blood cells, the population of mosquitoes, and counting of foreign bodies like a virus, bacteria, and parasite in a human body. Interpretations for critical points or equilibrium points have been given for network systems of differential equations associated with models. A practical method of applying these interpretations in administrating medicines to get control over diseases has been given. Order of priority in three types of critical points, namely, stable critical points, unstable critical points, and asymptotically stable critical points has been given. Conversions of differential equations of models into integral equations and applying Picard's iteration method to solve integral equations have been explained. A step-by-step approach has been used in designing models, solving models, and interpreting solutions of models for tropical diseases.

Chapter 5 focuses on Strongyloidiasis: biology, diagnosis and management of a most neglected tropical disease. Strongyloidiasis is a human parasitic ailment brought about by a whip-like nematode worm called Strongyloides stercoralis. Most humans get the infection by coming in contact with contaminated soil whereby the tiny worms penetrate the skin and enter the bloodstream from where it passes through the right side of the heart and lungs to mouth, stomach, and small intestines. Replication inside tainted people enables the disease to persevere for quite a long time. Strongyloides can cause a hyperinfection disorder which causes side-effects in numerous organ frameworks, including the central nervous system that can prompt death if untreated. The diagnosis is made by blood and stool tests. This chapter is intended to draw a more precise picture of the global prevalence, diagnosis and risk factors for S. stercoralis. The chapter also discusses the diagnostic approaches for detecting the infection, the morbidity caused and the recommended management. It further discusses some of the reasons why this infection is so neglected and the consequence of this for the estimated global prevalence.

Chapter 6 analyses the role of community participation in fighting schistosomiasis: lessons from Akka oases (Southern Morocco). In 2009, the Moroccan Ministry of Health in collaboration with the World Health Organization confirmed that the transmission of schistosomiasis had ceased in all previously endemic provinces of Morocco. This achievement couldn't have been achieved without the effective participation and engagement of local communities. In this chapter, studies of community participation strategies for schistosomiasis control in Akka oasis were reviewed. Southern Morocco was known to be a foci site for urinary schistosomiasis since 1937. Combined efforts between health officials and local community participation have achieved the elimination of this neglected tropical disease, where the last cases in these foci were recorded in the early 2000s. The results of this chapter can be used to increase awareness of the need for community participation in controlling neglected tropical diseases such as schistosomiasis.

Chapter 7 investigates the Geographical distribution of cutaneous leishmaniasis and its relationship with climate change in southeastern Morocco. The current chapter deals with one of the most neglected tropical diseases in Morocco, the cutaneous leishmaniasis. It is based on 10-year research (2010-2017) on the evolution of leishmaniasis taking climate change into account. Epidemiological and climatological data were collected from different administrations. The Geographic Information System (GIS) is chosen for interpolation, space-time analysis of climate data and map creation. The SPSS software was used for statistical analysis and to establish the relationship between Leishmaniasis and climatic conditions. Results show that the maximum number of cases is recorded in 2010 with 4407 people affected while the low number is recorded in 2014 with 18 cases. Results show also a clear link between climatic factors and the incidence of the disease. The distribution of the disease in the province is influenced by maximum temperature, aridity, and vegetation cover. Additionally, anthropogenic factors play a significant role in explaining the emergence or re-emergence of leishmaniasis in the region.

Ahmed Karmaoui
Southern Center for Culture and Sciences, Morocco

Acknowledgment

Climate change and anthropogenic impacts are expected to influence socio-economic and ecological systems, including health. In this book, the relationship between climate, anthropogenic factors, and health, including the neglected tropical diseases (NTDs) has been explored. In this collective book, a multidisciplinary collection of chapters was produced. Each one gives insight or detailed information on one or several diseases in many fields, such as epidemiology, medicine, and ecology. Using different approaches such as case studies, reviews, and modeling methods original results were explored.

The book has seven chapters that are organized into three sections. The first section discusses the associations between climate change, humans, animals, plants, and health and section two highlights the climate change and neglected tropical diseases associations. The third section explores the climate change and anthropogenic impacts on some tropical neglected diseases from two vulnerable continents, Africa and Asia.

This collective book is a result of the collaboration of three editors and several authors, researchers, and academics, with the valuable support of the IGI global team over a period of 10 months.

Many thanks go to the authors for their chapters and to reviewers for their effort to improve the content and the form of the seven selected chapters.

I am grateful for the collaboration with my colleagues, the co-editors Prof. Abdelkrim Ben Salem and Prof. Ashfaq Ahmad Shah.

Special thanks go to my little family, my wife, Siham; my two children, Yassin and Amine. Many thanks are addressed to my dad, Mimoun, for his continuous encouragement.

Section 1

Climate Change and Humans, Animals, and Plants Health Associations

Chapter 1
Climate Change and Health Impacts in Pakistan

Saddam Hussain
National University of Sciences and Technology, Pakistan

Sobia Siddique
National University of Sciences and Technology, Pakistan

Ashfaq Ahmad Shah
https://orcid.org/0000-0001-9142-2441
National University of Sciences and Technology, Pakistan

ABSTRACT

Conferring to the Global Risk Index, Pakistan is ranked as the 7th most susceptible country to the inexorable influence of climate change. Before this century ends, the annual mean temperature in Pakistan is expected to rise from 3°C to 5°C for a focal worldwide discharge situation. Usually, annual precipitation is not relied upon to have a critical long haul pattern. Ocean level is relied upon to ascend further by 60 centimeters. All these climatic events are likely to disrupt the economy, lives, and the socio-political aspects of human life. Pakistan has already witnessed massive loss in terms of human, infrastructural, and economic aspects. The chapter is designed to understand both the direct and indirect health risks associated with frequent climatic events like floods, drought, and heat waves in Pakistan. After analyzing the available literature, it was observed that floods and drought have direct and indirect health risks associated with them while in case of heat waves, health risks cannot be established precisely as multiple variables are involved, playing a significant role.

DOI: 10.4018/978-1-7998-2197-7.ch001

INTRODUCTION

The perils of climate change are coming out of shadows and manifesting themselves in various environmental phenomenon or climatic events. Before listing out these climatic events, it is imperative to define 'climate change' in a way that a layman is able tounderstand it. Therefore, two important definitions of climate change are adopted by the international bodies working on climate change namely the Intergovernmental Panel on Climate Change (IPCC) and United Nations Framework Convention on Climate Change UNFCCC. IPCC demarcated climate variation as "change in the state of the climate that can be identified (e.g. using statistical tests) by changes in the mean and/or the variability of its properties, and that persists for an extended period, typically decades or longer". While UNFCCC's definition of climate change is "change of climate which is attributed directly or indirectly to human activity that alters the composition of the global atmosphere and which is in addition to natural climate variability observed over comparable time periods" (IPCC, 2007a).

Climate-induced changes may be intense and sudden, or slow-onset. The latter includes changes such as soil dilapidation and deterioration in farming productivity, water scarcity, decline in biodiversity, dilapidation of ecosystem services while sudden climatic changes can be categorized into life-threatening hazards like drought, floods, intense rainfall, Glacial Lake Outburst Floods (GLOF) or continuous hot spell. (Hoermann et al., 2010). In its Fourth Assessment Report, the IPCC dissipated a lot of suspicionwith respect to climate change. Global warming is being experienced without a doubt and scientists are 95 per cent sure that anthropogenic activities are the main drivers of climate change (IPCC, 2014). Scientists from various parts of the globe agree that the contemporary rate of warming around the world has the fastest trend when compared with trends in the past. IPCC's 4th Assessment Report provided unambiguously thorough prognoses for the current century and showed that the contemporary rate of global warming is unlikely to slow down but it is expected to accelerate abruptly. The best available estimate for global warming is 3°C rise by end of the current century. This rise in average temperature worldwide will have dire consequences for all living beings on earth. The entire human civilization will suffer from the burn of climate change (IPCC, 2007).

By looking from the spectrum of human health, climate change is expected to distress the fundamental elements of health: food security, cleaner water and air. Numerous diseases are profoundly climate or environmentally delicate in the sense that they are more sensitive to various climatic events. For instance, greater variability in the climate means expanding air toxins which can, in turn, increase the chances of airborne and respiratory ailments. As dangers from unexpected climate events, for example, floods and violent winds become more visiting and annihilating, the transmission of sicknesses through polluted water and sullied rations is likewise prone to multiply enormously. Climate change is also sabotaging the progress that fights against the climate-sensitive diseases.

Effects of climate change can be broadly seen. It can be classified into direct and indirect impacts when talking in the context of anthropological well-being. The direct human health impacts inmate from floods, storm surges, tropical cyclones and heatwaves. As global warming is mounting up, heat waves are predicted to be more ubiquitous with elongated duration leading to an increase in cerebrovascular, respiratory and cardiovascular diseases (Hales, Edwards, and Kovats 2003). Heat-strokes are more prevalent among outdoor labourers and poor people in parched and semi-parched regions (Chaudhury, Gore, and Ray 2000). Deaths due to heat wave have been found especially among the vulberable communities: poor people, the old, and daily wage workers that include rickshaw drivers and agricultural labourers (Lal 2003).

In South Asia, greater rainfall variability is anticipated to escalate the recurrence rate of devastating floods (Cruz et al. 2007). These changing patterns of rainfall are witnessed in many countries. For instance, high-intensity monsoon rainfall in Nepal, Bangladesh and India resulted in floods that lead to the demise of 2000 individuals and dislodging of 20 million people (WHO 2005). Flooding has been related to immediate and prompted well-being dangers. Direct dangers incorporate demise due to suffocating and injury from being hit by articles moving quickly in the flood stream (Ahern et al. 2005). Frequently, the paramount damage comes from prompted impacts. Flooding makes a domain helpful in the transmission of difficulty. On the off chance, floodwater becomes debased with human or creature waste which can increase the rate of faecal-oral ailment transmission, permitting diarrheal ailment and other bacterial and viral diseases. Faecal-oral transmission of illnesses is worrisome in areas where limited access is available to clean water and sanitation. Increments in diarrheal illness, cholera, loose bowels, and typhoid are of explicit concern (Morgan et al. 2005). Similarly, cyclones also pose significant threats to human health. The common penalties associated with cyclones are sinking and the spread of contagious maladies. Many deaths have been witnessed due to drowning during cyclones (Keim 2006). Chronic or ceaseless illness are likely to increase after high-intensity cyclones, as the victims experience physical weakness, mental stress and have constrained coping capacities (Hess, Malilay, and Parkinson 2008).

Multiple diseases such as cryptosporidiosis, giardiasis, salmonellosis, cholera and vector-borne diseases especially malaria is caused due to contaminated water which spreads catastrophically in warmer climate (Hales, Edwards, and Kovats 2003). In South Asian region, the most common disease associated with water is diarrhea, which is one of the leading factors of mortality and morbidity in the region (Zaidi, Awasthi, and de Silva 2004). Lack of proper sanitation, availability of freshwater and consummation of unhygienic water are attributable to the spread of diarrhoea (Ezzati et al. 2004). In near future, these factors will be further enhance climate change (Cruz et al. 2007). Likewise, rise in the atmospheric temperature and precipitation are responsible for spreading vector-borne diseases. Most common example of such disease is malaria across the arid and temperate parts (Hales, Edwards, and Kovats 2003).

On the authority of Global Climate Risk Index of 2017, Pakistan is graded as the seventh most exposed country to the inexorable effects of climate change. Therefore, this chapter will be discussing about the influences of climate change on anthropological health in Pakistan.

METHOD

Study Area

According to the Global Change Impact Studies Centre (2017): "Between 1997 and 2016, Pakistan lost an average of 523 people. 1 lives per year i.e. 10,462 live in 20 years, which makes a total of3.27 lives per million. Pakistan was ranked 4th in terms of property damage and the largest contribution to these damaged numbers came from the 2010 floods. Besides this, Pakistan has suffered from prolonged droughts (1998-2002, 2014-17), heat waves (2011, 2014), the 2014 cyclone Nilofer, and Glacial Outburst Lake Floods (GLOF) events". This study helps to understand the health risks associated with different climatic events in different areas of the country. For example, health risks associated with floods were studied across the flood-affected areas in Pakistan, while drought associated health risks were studied in the two provinces of Pakistan namely Sindh and Baluchistan. While, health risk associated with heat-waves were studied in the metropolitan city of Karachi.

Methodology

In case of this research study, methodology is constructed on the basis of "library research" which is one of the five research strategies according to "DOT-Framework Identities". Library research refers to a study devised to obtain as a summary of prevailing or existing work (for instance, literature review, completion analysis etc.) therefore, previous studies conducted on the topic of interest were consulted and analyzed in order to extract and scrutinize relevant information on the health risks associated with natural hazards. In this regard, numerous reports published by governmental institutions, international and local organizations were compiled and studied on three prevailing natural hazards for instance heat-waves, drought and floods.

Floods and Human Health

Pakistan is bestowed with assorted climate and diverse topography. Its climate is often dubbed as hot and dry, but over the recent years, substantial variations were detected in the climate. Fast glacial retreat in the mountainous ranges and hefty precipitation in the retention basin or areas look for the Indus river and its major offshoots i.e., Kabul, Ravi, Sutlej, Chenab, Swat and Beas and other minor tributaries. The stream in the waterways, which boost Pakistan's economy, from time to time transform into ruinous floods and result in substantial harms to open and private area properties other than loss of valuable human lives. High floods are in some cases brought about by the development of transitory normal dams through avalanche or icy mass developments and their resulting breakdown (Government of Pakistan, 2017).

Numerous regions and urban concentrations situated along the rivers or waterways of banks are an extraordinary hazard that cause various forms of flood. For example, streak or flash flood, riverine and town floods, especially in Punjab and Sindh regions. These floods cause harm to thousand acres of land rearing fruitful rural grounds, standing harvests and side by side influence the connection of human concentrations with money related misfortunes. Most important and immediate floods harm is inflicted on farming grounds, harvesting yields, city and rustic shanties plus, other cloistered and open property (Government of Pakistan, 2017). Floods pretty much destroy extensive land territory/standing harvests and influence and dislodge massive number of individuals. It is the most regular common danger being looked at by the nation since its creation in 1947. Because of an Earth-wide temperature boost and fast environmental change, recurrence of floods has expanded.

Attributable to antagonistic effects of environmental change in the ongoing years, vulnerability of networks and coastal flooding has expanded additionally. Flood harm is mostly caused due to riverine flooding in fundamental waterways and blaze floods in Secondary and Tertiary Rivers/Hill Torrents, Coastal flooding because of Cyclone and urban flooding because of heavy rains and deficient drainage waste evacuation services in addition to GLOFs in northern areas of the country. During the past sixty-seven years, the absolute misfortunes that credited to significant flood occasions in Pakistan exceeded US$ 38.171 billion while almost 12,502 individuals lost their lives out of which half of the all-out direct misfortunes in history had occured in the last 6-7 years.

Initially, floods result in physical or infrastructural destruction but as the flooded water withdraws, more dangers can be witnessed to the well-being and prosperity of individuals. Flood crisis in Pakistan is tremendously affecting the prosperity of considerable number of people and the ability to pass on emergency and routine health care or social human services to the effected population (Jongman, 2015). Floods are recognized with an extended risk for diarrheal sicknesses. A few examinations have shown that these diarrheal contaminations can increment in weeks or months after floods, both in developed and undeveloped nations.

According to a report by the United Nations Development Program Bureau for Crisis Prevention and Recover, Pakistan is ranked as the fourth defenceless country against flood related calamities in South Asia. The positioning is dependent on Disaster Risk Index which is based on the mortality-based record and appearance of individuals to the risky circumstances.

Floods are most importantly a wellbeing emergency. They pass through suffocating and lethal wounds, debase clean water supplies and give rise to vector-borne ailments. As per the insight of Pakistan's government, floods influenced 20,184,550 individuals and brought about 1,985 passings and 2,946 wounds (Solberg, 2010). In the regions influenced by floods, pestilence illness episode is normal. In the annual report published by UNICEF (2014) after 2012 floods, 8046 instances of measles were accounted for from PJETS. Moreover, the report says that around 480,000 kids were treated for serious loose bowels, pneumonia and some intense respiratory contaminations. Likewise, 77000 kids were determined to have lack of healthy sustenance in the influenced territory. The most common inescapable diseases in the flood-affected areas are categorized by Shabir (2013) in his study are Tuberculosis, Diarrhea, Hepatitis C and, acute respiratory infection, measles and vector born malady such as hemorrhagic fever and malaria. The Floods profoundly affect Malaria peril and require preventive and remedial measures. Considering the annual cases certified in the year 2010, it was found that there were as high as 1 million cases of Annual Parasite Incidence (UNICEF, 2014).

Floods have obstructed 599,459 sq. kilometer of landscape, loss of 11,239 lives, caused mishaps that costed over PKR 39 billion to the countrywide economy and left 180,234 towns submerged. According to the Asian Development bank, Pakistan has encountered 21 floods from 1950-2011 that have expected financial harm of $19 billion and subsequently preceding three back to back years. Pakistan observed two additional floods accounting more to the misfortune (Ali, 2013). Latest reports published by local authorities explicitly mentioned the death cases of 83 unfortunate individuals who lost their lives to flooding. Among these there were 22 cases in AJ&K, 25 in KP, 23 in Punjab and 10 in Balochistan (Govt, 2019). Likewise, the flash flood was reported in district Neelam, killing 24 and injuring 7 people (OCHA,2019).

Drought and Human Health

Drought or dry spell has numerous aspects. It might arrive at different degrees of seriousness and can show up in any area around the world, with an effect extending from slight individual burden to imperiled nation-state. The methodology by different examiners varies as indicated by their specific advantages and foundations. It is not startling that a widespread room of meanings has been created. Numerous descriptions have been proposed to move towards the investigation of the dry season or drought with a specific goal in mind (WMO, 1975). Many are altered forms of the old definitions and their consideration here is mostly for the sake of keeping a record.

Drought contrasts from different meteorological wonders are in chronological perspectives. Its commencement and expiry are frequently somewhat equivocal as for time. Its period might be usually long. Wide-range of definitions show that they can be characterized by the criteria utilized such as wind, evapotranspiration, air temperature, humidity, streamflow condition and precipitation might be utilized either in a mix or alone as a variable(s). A few explanations determine a transient control while others are autonomous of time. The dry season might have a hurtful impact both on naturally or engineered vegetation and living creatures (Warwick, 1975). Drought has been wrongly considered as an irregular and uncommon scene even though it is an intermittent and ordinary climatic element. It might happen in various climatic zones all over the world. There are huge local varieties in the qualities of dry season. Significant lot of beneath ordinary precipitation brings about dry season; basically expressed, they are total dangers. An impermanent precipitation variation is called dry season. Infact, low precipitation is a perpetual element of the atmosphere of parched locales (University of Florida, 1998). The dry season is a risky peril. The dry spell must be viewed as with respect to average conditions over a time of 3 decades of wonders identified with evapotranspiration, temperature, precipitation, relative wind speed and humidity which are considered 'ordinary' when compared with discrete meteorological posts. Dry spell fallouts when noteworthy can lead to temperature peculiarity over an interminable timespan; bringing about exorbitant evaporation, postponements at the beginning of the blustery season or awkward precipitations for main harvest development phases. Truncated comparative dampness and fast wind speed further causes inconvenience in the circumstances (UOF, 1998).

According to the UNCCD, drought can be demarcated as a "phenomenon occurring due to below-normal precipitation, which adversely affects land resource production systems when precipitation has been signed in hydrological imbalance, culminating in unfavourable production of the land resource systems". There are factors such as evaporation, soil types, level of underground water reservoirs and vegetation that are closely associated with precipitation and play a major role in the manifestation of drought or dry season. Thinking about the variables, four dry spells or drought are generally recognized. These include agricultural, hydrological, meteorological and socio-economic drought. The meteorological dry spell is the result of continued shortfall of precipitation; the hydrological dry spell is related to the absence of watercourse and agriculture or agrarian dry season is perceived by soil dampness shortage and socio-economic droughts occurs when water demand exceeds the supply (Keyantash, et al., 2002).

In its territory, Pakistan's government-issued warning for drought in multiple districts of Gilgit-Baltistan, Sindh, Baluchistan and Punjab province. The most seriously influenced areas come under the territory of Sindh province, for instance, Dadu, Umerkot, Tharparkar, Thatta, Kambar Shahdadkot and Sanghar, altogether these areas have been proclaimed "calamity-affected". The total number of villages in these areas are roughly 315 but the approximations for affected populace are not yet accessible (OCHA, 2018). Likewise, after the quick-dry spell need evaluation, the legislature of Balochistan announced that the 18 most noticeably terrible dry season influenced regions are Killa Abdullah, Pishin, Chaghi, Noshki, Washuk, Kharan, Kech, Panjgur, Awaran, Jhal Magsi, Jaffarabad, Zhob, Quetta, Barkhan, Kohlu, Killah Saifullah, Kachhi and Naseer Abad. (IFRC, 28 Jan 2019). In Baluchistan, around 48% of the population is living below the poverty line and rural poverty is 51% of the total poverty which is extremely high. Estimations for food-insecure households are also astoundingly high. Among the 20 most food-insecure districts of Pakistan, 10 lie in Baluchistan province alone. It has been identified that due to prevailing drought-like conditions, not even a single district in the province can be saved from the severity of food insecurity (SDPI et al., 2009).

Rising drought situations are distressing somewhere in the range of 5,000,000 individuals in Balochistan and Sindh provinces. With each passing day, drought conditions are worsening in the affected locale where more than 71 percent of the family units are recognized as food insecure and the incidence of malnourishment is augmented by 30 percent (OCHA, 2019).

In Sindh, more than 3700 leishmaniasis cases were accounted for in 2018. Also, there is continuous flare-ups of Crimean Congo hemorrhagic fever (CCHF), malaria, leishmaniasis and chikungunya and leishmaniasis. All these diseases fall in the category of vector-borne illness. The staggering number of leishmaniasis cases were reported in one district of Baluchistan in 2018. It is significant that with the continuous dry season, the number of agents for these vectors in these territories will increase. This will prompt spread of the outbreaks. Children and women are the most seriously influenced groups effected by the dry season in Baluchistan. Around 35% of the deaths among yound children in Pakistan are caused due to to undernourishment and around 60% are caused by poor sanitation and water-borne maladies (UNICEF 2011). The effect of the rations deficiency during catastrophes is intense on children and women. Baluchistan has the most astounding maternal and infant child death rate. About 130 out of 1,000 children conceived, bite dust in the territory (PDMA, undated).

The dry season circumstances in Sindh and Balochistan are quickly forming one of the most terrible fiascos in Pakistan. Sindh and upland Baluchistan in the south are most intensely influenced by the extreme dry season. These influenced territories of Sindh and Balochistan have inclined to water deficiencies, precipitation estimated in the course of the most recent couple of years has arrived at a record low, with negligible or no precipitation. PMD posited in its warning that "serious to extreme drought-like conditions" have developed over a large portion of the southern pieces of Pakistan because of the absence of summer downpour. The unexpected decrease in precipitation in majority of the upland regions has resulted in diminishing of water yield from springs and tube wells. On Feb 07, 2019, NDMA sorted out a Strategic Coordination Forum/Steering Committee Meeting, upheld by UNOCHA with the motivation to talk about the dry spell reaction plan 2019, and the Chairman of NDMA authoritatively mentioned universal partners to help the significant issue and dry spell unfortunate casualties, including IFRC (IFRC, 2019). An itemized needs appraisal in the month of April 2019 was carried out by Pakistan Red Crescent Society (PRCS) to distinguish high hazard or risk and need zones of Baluchistan and Sindh. Attributable to low rainfall, the underground water level is sinking and bringing about evaporation of water springs amaking it difficult for the public to get water for fulfilling their basic needs and and livestock use. It additionally influences the cleanliness and sanitation in the civic spaces. This has triggered the broad lack of healthy sustenance identified among children and women by FAO/WFP (IFRC, 2019).

One of the most exceedingly terrible dry seasons in Pakistan's history has activated "alarmingly high" levels of unhealthiness and trouble in the hardest-hit zones where individuals have been compelled to drink salty or debased water, the Red Cross said on Friday. Children and women, particularly pregnant or breastfeeding, are especially powerless against vomiting, diarrhoea and fever expedited by an absence of safe water in the most noticeably awful influenced regions of Southern Baluchistan and Sindh (Dawn, 2019).

Upon the invitation of Pakistan, the UN and different accomplices are doing a drought appraisal in eight locales of Sindh Province. The drought assessment will be carried out by the established National Disaster Consortium in the supervisor of the International Organization for Migration (IOM). Other renowned Universal Official Organizations such as the Food and Agriculture Organization (FAO) and World Food Program (WFP) are significant supporters of this assessment because of their settled field nearness in the focused-on locale (WFP, 2018). The authority has announced condition of catastrophe in drought influenced territories in the regions of Khairpur, Sukkur, Umerkot, Sanghar, Tharparkar, Badin, Thatta, Kachho, Jamshoro and Dadu because of not foreseeing rainfall during monsoon season. Local government began disseminating alleviation bundles to influenced families. The dispersion is required to last around a quarter of a year (NDMA, 2018).

The Pakistani Government has provided 2.5 million individuals with emergency health care, cash disbursements and nutritional services. International developmental organizations are also supporting the Government by diverting current projects and assets to need areas. There is an earnest requirement for extra sustenance help, access to safe drinking water, nourishment administrations, social insurance, and employment support (OCHA, 2019).

Heatwaves and Human Health

In South Asia, heat-waves are a prevailing climatic danger with perceivable effects on human life and natural ecosystems. Over the past decade, it was recorded that the recurrence of hot days was three times more than the recurrence of cold days being recorded at the universal level. An examination by Pakistan Meteorological Department (PMD) indicated that the occurrence of warm wave days has increased significantly. The sweltering climate is a prerequisite for heatwaves.

A heat wave comprises of an all-inclusive stretch of scorching climate instead of the normal conditions of a territory at a particular season. WMO characterizes heat wave as "when the everyday most extreme temperature in excess of five sequential days surpasses the normal greatest temperature by 5 °C, from the ordinary temperature of a territory". Heat waves occur regularaly in the pre monsoon months (May and June) in the parched and semi-parched areas of Pakistan where the number of days and surrounding warmth differs from area to area. According to Steffen, heat waves have a few noteworthy attributes. These include "(i) the number of heatwave days and the annual number of summer heatwave occasions; (ii) the length of the longest heatwave in a season; (iii) the normal abundance temperature expected during a heatwave and the most sultry day of a heatwave; and (iv) the event of the principal heat-wave occasion in a season". These attributes are imperative to comprehend for recognizable proof and more profound investigation of the idea of warm waves.

Considering a meteorological opinion, heat-wave instances can be divided into two more extensive sort of moist and dry heat waves. Dry heat waves are described by dry climate condition with cloudless sky and enormous contributions of sun radiation. It might be joined by blustery conditions which can lead to the expansion of warmth stress. Dry heat waves occur in areas comprising of mainland or areas with Mediterranean atmosphere where air is warm. Conversely, moist heat waves are portrayed by extremely warm, severe, sticky conditions for the duration of the day and night. Regularly with night-time overcast spread, a component that averts loss of heat gathered for the duration of the day and along these lines gives little help in the evening time. Such heatwaves are regularly a component of mid-scope mild and sea atmosphere which might be common to certain locales (Hunt, 2007). In view of these qualities, heat waves are bound to occur in areas that have an exceptional factor summer atmosphere or a reasonable sweltering season and, as needs be, may result from a scope of huge scale meteorological circumstances and atmosphere related robotic procedures (Chang and Wallace). Areas with an unmistakable sweltering seasons are not resistant from warm waves. Unordinary mixes of land, sea and air conditions may give climatological setting to momentary atmospheric shocks and the event of outrageous temperature and stickiness. The planning of heatwave occasions might be identified with the general atmospheric conditions. For instance, unfortunate heatwave occasions in southern Asia seem to happen right off the bat in the mid-year to preceding the entry of the late spring rainstorm. One approach to check the connection between atmosphere and the event of heatwave occasions is to consider the atmosphere types as per the standard arrangement of atmosphere related with heatwave occasions which have had noteworthy societal effects (WMO, 2015). The lamentable heatwave is not effective for multiple days with extremely high temperature, yet it is a combined effect of moistness and temperature. To evaluate this consolidated impact, Heat Index is created. This index explains the heatwave condition appropriately on a specific day.

Heatstroke is a restorative crisis and is a type of hyperthermia under which body temperature rises drastically and can act deadly if appropriate treatment is not provided in time. The body's temperature rises quickly, the perspiring system falls flat, and the body cannot chill off thus the body temperature can rise to 106 degrees Fahrenheit or higher in a matter of few minutes that is 10-15 minutes. Human body produces heat because of digestion and disseminates the warmth by its radiation through the skin or by the dissipation of perspiration. It case of outrageous warmth, high stickiness or incredible physical effort under the sun, the body will be unable to adequately disseminate the warmth resulting in an increase in body temperature. Another reason for heatstroke is parchedness. A dried-out individual would be unable to sweat quick enough to scatter heat which makes the body temperature rise. Regular signs and side effects of heatstroke incorporate plentiful perspiring or absence of perspiring, with either hot red or flushed dry skin, shortcoming/torpidity, cerebral pain, increased body temperature, mind flights, disarray/dazedness and slurred discourse. Heatstroke can cause demise or lasting organ harm any other disabilities if not treated appropriately in time.

While all segments of the populace are in danger of ailment and demise when presented to expanded balminess, and particularly outrageous heat, specific sub-bunches are more defenceless than others. Be that as it may, connections among temperature and health of wellbeing, effects are neither uniform nor banal which are affected by various complex and cooperating components including environmental, medical, biological, social and topographical elements.

The older segment of the population seems more susceptible than younger population to encounter sickness and in the worst case, death because of outrageous warmth exhibited in various areas, continents and across the biosphere. Few investigations demonstrate no immediate connection between heat-related deaths or ailment and age, proposing social variables might be of impact at a local scale. Heat-related mortality likewise gives off an impression of being related with a scope of previous incessant ailments conditions, including cerebrovascular, cardiovascular, endocrine, respiratory, sensory system conditions genitourinary and emotional wellness issue (Fouillet et al., 2006, Haines et al., 2006), Other distinguished sub-bunches with expanded weakness to heat-related sickness incorporate those working outside or in extremely warm conditions and those utilizing specific prescriptions. Besides, populace living in locales who previously encountered sweltering climate may be able to find temperature increase non-survivable, making their current vulnerabilities pressing to address.

In Pakistan, the most extreme cases of heat strokes were witnessed in the city of Karachi in 2015, where heatwaves resulted in the demise of almost 1200 lives creating serious ailments among the population due to the prevailing heat stress. Measuring the impact of heat-waves was made possible by instrumentalizing a tool called heat index. The said index combines the moistness, direction and pressure of air with aggregate temperature and wind rate in order to measure the impact of heatwaves on human beings residing in a particular area. On the scale of heat index, maximum temperature recorded was 48.8 °C. But heat-wave index recorded on June 20, 2015 was found to be 66 °C due to diminished air pressure, wind speed and mounting humidity. The sufferers were generally aged people, women, children and particularly labourers. Major reasons for the loss of human lives were dehydration, heat exhaustion and heat cramps (Chaudhry, et al., 2015).

CONCLUSION

The health risks associated with natural hazards can be lethal for human beings in case of Pakistan as the country witnessed human losses to various health hazards posed by various climatic events. Floods stand alone when it comes to the greatest health threat across the country. Over the last 67 years, floods have resulted in loss of aroud 12,502 human lives and acted as a stimulus in the spread of various life-threating diseases such as measles, respiratory infections, pneumonia Tuberculosis, Diarrhea, Hepatitis C vector born malady such as hemorrhagic fever and malaria. Likewise, health risks associated with drought are flaring up in Sindh and Baluchistan.

In 2018, Sindh alone witnessed 3,700 incidents of dengue fever concomitant with other life-threating diseases such as chikungunya, Crimean Congo hemorrhagic fever (CCHF), leishmaniasis and malaria. Similarly, more than 5,000 cases of leishmaniasis were witnessed in Baluchistan in the drought tainted region of Kila Abdullah. Most extreme cases of heat strokes were witnessed in the city of Karachi in 2015, where heatwaves resulted in the demise of almost 1200 lives creating serious ailments among the population due to the prevailing heat stress. Therefore, it is valid to say that in the upcoming years, if government remains heedless to the health risks associated with natural hazards, the repercussions will be more severe and frequent.

REFERENCES

Ahern, M., Kovats, R. S., Wilkinson, P., Few, R., & Matthies, F. (2005). Global health impacts of floods: Epidemiologic evidence. *Epidemiologic Reviews*, *27*(1), 36–46. doi:10.1093/epirev/mxi004 PMID:15958425

Bai, L., Ding, G., Gu, S., Bi, P., Su, B., Qin, D., ... Liu, Q. (2014). The effects of summer temperature and heat waves on heat-related illness in a coastal city of China, 2011–2013. *Environmental Research*, *132*, 212–219. doi:10.1016/j.envres.2014.04.002 PMID:24815333

Beggs, P. J. (2000). Impacts of climate and climate change on medications and human health. *Australian and New Zealand Journal of Public Health*, *24*(6), 630–632. doi:10.1111/j.1467-842X.2000.tb00531.x PMID:11215015

Bustinza, R., Lebel, G., Gosselin, P., Bélanger, D., & Chebana, F. (2013). Health impacts of the July 2010 heat wave in Quebec, Canada. *BMC Public Health*, *13*(1), 56. doi:10.1186/1471-2458-13-56 PMID:23336593

Cerutti, B., Tereanu, C., Domenighetti, G., Cantoni, E., Gaia, M., Bolgiani, I., ... Cassis, I. (2006). Temperature related mortality and ambulance service interventions during the heat waves of 2003 in Ticino (Switzerland). *Sozial- und Praventivmedizin*, *51*(4), 185–193. doi:10.100700038-006-0026-z PMID:17193780

Chang, F. C., & Wallace, J. M. (1987). Meteorological conditions during heat waves and droughts in the United States Great Plains. *Monthly Weather Review*, *115*(7), 1253–1269. doi:10.1175/1520-0493(1987)115<1253:MCDHWA>2.0.CO;2

Chaudhry, Q. Z., Rasul, G., Kamal, A., Mangrio, M. A., & Mahmood, S. (2015). Task Force/ Fact Finding Mission by GOP on "Causes of Severe Heat Wave in Karachi. Technical Report, 2015. Coley, D.

Chaudhury, S. K., Gore, J. M., & Ray, K. S. (2000). Impact of heat waves over India. *Current Science*, *79*(2), 153–155.

Choi, J., & Meentemeyer, V. (2002). Climatology of persistent positive temperature anomalies for the contiguous United States (1950–1995). *Physical Geography*, *23*(3), 175–195. doi:10.2747/0272-3646.23.3.175

Cruz, R. V., Harasawa, H., Lal, M., Wu, S., Anokhin, Y., Punsalmaa, B., . . . Huu Nin, N. (2007). Asia. In Climate Change 2007: Impacts, Adaptation and Vulnerability. Contribution of Working Group II to the Fourth Assessment Report of the Intergovernmental Panel on Climate Change.

Dalip, J., Phillips, G. A., Jelinek, G. A., & Weiland, T. J. (2015). Can the elderly handle the heat? A retrospective case-control study of the impact of heat waves on older patients attending an inner-city Australian emergency department. *Asia-Pacific Journal of Public Health*, *27*(2), NP1837–NP1846. doi:10.1177/1010539512466428 PMID:23212171

Dawn. (2019, March 2). *No TitHunger, disease "alarmingly high" in drought-hit Pakistan: Red Crossle.* Retrieved from https://www.dawn.com/news/1467107/hunger-disease-alarmingly-high-in-drought-hit-pakistan-red-cross

Ezzati, M., Lopez, A. D., Rodgers, A. A., & Murray, C. J. (2004). *Comparative quantification of health risks: the global and regional burden of disease attributable to selected major risk factors.* World Health Organization.

Fouillet, A., Rey, G., Laurent, F., Pavillon, G., Bellec, S., Guihenneuc-Jouyaux, C., ... Hémon, D. (2006). Excess mortality related to the August 2003 heat wave in France. *International Archives of Occupational and Environmental Health*, *80*(1), 16–24. doi:10.100700420-006-0089-4 PMID:16523319

Fouillet, A., Rey, G., Laurent, F., Pavillon, G., Bellec, S., Guihenneuc-Jouyaux, C., ... Hémon, D. (2006). Excess mortality related to the August 2003 heat wave in France. *International Archives of Occupational and Environmental Health*, *80*(1), 16–24. doi:10.100700420-006-0089-4 PMID:16523319

Government of Pakistan. (2017). *Annual Flood Report 2017.*

Grum, R. H. (2011). The central European and Russian heat event of July–August 2010. *Bulletin of the American Meteorological Society*, *92*(10), 1285–1286. doi:10.1175/2011BAMS3174.1

Haines, A., Kovats, R. S., Campbell-Lendrum, D., & Corvalán, C. (2006). Climate change and human health: Impacts, vulnerability, and mitigation. *Lancet, 367*(9528), 2101–2109. doi:10.1016/S0140-6736(06)68933-2 PMID:16798393

Hales, S., Edwards, S. J., & Kovats, R. S. (2003). Impacts on health of climate extremes. *Climate change and human health: Risks and responses*, 79-96.

Hess, J. J., Malilay, J. N., & Parkinson, A. J. (2008). Climate change: The importance of place. *American Journal of Preventive Medicine, 35*(5), 468–478. doi:10.1016/j.amepre.2008.08.024 PMID:18929973

Hoermann, B., Banerjee, S., & Kollmair, M. (2010). *Labour migration for development in the western Hindu Kush-Himalayas: Understanding a livelihood strategy in the context of socioeconomic and environmental change*. Kathmandu: International Centre for Integrated Mountain Development; Available at http://lib.icimod.org/record/8050/file s/attachment_695.pdf

Hunt, B. G. (2007). A climatology of heat waves from a multimillennial simulation. *Journal of Climate, 20*(15), 3802–3821. doi:10.1175/JCLI4224.1

Im, E. S., Pal, J. S., & Eltahir, E. A. (2017). Deadly heat waves projected in the densely populated agricultural regions of South Asia. *Science Advances, 3*(8). doi:10.1126ciadv.1603322 PMID:28782036

IPCC. (2007). Climate change 2007: Synthesis report. Geneva, World Meteorological Organization (WMO), United Nations Environmental Program (UNEP).

IPCC. (2007a). [*The physical science basis*. Geneva: Intergovernmental Panel on Climate Change. References]. *Climatic Change*, 2007.

IPCC. (2014). [Impacts, Adaptation and Vulnerability, Fourth Assessment Report.]. *Climatic Change*, 2007.

IPCC. (2014). Impacts, Adaptation and Vulnerability, Fourth Assessment Report. *Climatic Change*, 2007.

Jongman, B., Winsemius, H. C., Aerts, J. C., de Perez, E. C., van Aalst, M. K., Kron, W., & Ward, P. J. (2015). Declining vulnerability to river floods and the global benefits of adaptation. *Proceedings of the National Academy of Sciences of the United States of America, 112*(18), E2271–E2280.

Keyantash, J., & Dracup, A. J. (2002). The Quantification of Drought: An Evaluation of Drought Indices. Bulletin of the American Meteorological Society. 83. 1167-1180. <1191: TQODAE>2.3.CO;2Dawn. (2019, March 2). *No TitHunger, disease "alarmingly high" in drought-hit Pakistan: Red Crossle*. Retrieved from https://www.dawn.com/news/1467107/hunger-disease-alarmingly-high-in-drought-hit-pakistan-red-cross doi:10.1175/1520-0477(2002)083

Klinenberg, E. (2015). *Heat wave: A social autopsy of disaster in Chicago*. University of Chicago Press. doi:10.7208/chicago/9780226276212.001.0001

Lal, R. (2003). Soil erosion and the global carbon budget. *Environment International, 29*(4), 437–450. doi:10.1016/S0160-4120(02)00192-7 PMID:12705941

Memon, N (2012). Malevolent Floods of Pakistan 2010-2012.

Morgan, J. (2005). Global trends in candidemia: A review of reports from 1995–2005. *Current Infectious Disease Reports, 7*(6), 429–439. doi:10.100711908-005-0044-7 PMID:16225780

NDMA. (2018). *Report on Prevailing Drought Like Situation in Sindh With*. (November). Retrieved from https://reliefweb.int/sites/reliefweb.int/files/resources/Drought Situation Report of Sindh Particularly in District Tharparkar-2018%2812-11-2018%29.pdf

NDMA. (2018). *Report on Prevailing Drought Like Situation in Sindh With*. (November). Retrieved from https://reliefweb.int/sites/reliefweb.int/files/resources/Drought Situation Report of Sindh Particularly in District Tharparkar-2018%2812-11-2018%29.pdf

Nishtar S. (2007). Health Indicators of Pakistan – Gateway Paper II. Islamabad, Pakistan: Heartfile; 2007.UNICEF Pakistan Situation Report No. 21, Final Report on 2012 Monsoon Floods in Pakistan, April 7, 2013, and Humanitarian Action for Children (HAC) 2014 Report.

OCHA. (2019). *Asia and the Pacific: Weekly Regional Humanitarian Snapshot (September 25 - October 1, 2018) No Title*. Retrieved from https://reliefweb.int/sites/reliefweb.int/files/resources/ROAP_Snapshot_190121.pdf

Pakistan Health Cluster - Floods in Pakistan. (2010). (*No. 19, Focus on donors, September 28, 2010*). Cairo: World Health Organization Regional Office for the Eastern Mediterranean.

Palecki, M. A., Changnon, S. A., & Kunkel, K. E. (2001). The nature and impacts of the July 1999 heat wave in the Midwestern United States: Learning from the lessons of 1995. *Bulletin of the American Meteorological Society*, *82*(7), 1353–1367. doi:10.1175/1520-0477(2001)082<1353:TNAIOT>2.3.CO;2

Pezza, A. B., van Rensch, P., & Cai, W. (2012). Severe heat waves in Southern Australia: Synoptic climatology and large-scale connections. *Climate Dynamics*, *38*(1–2), 209–224. doi:10.100700382-011-1016-2

Schaffer, A., Muscatello, D., Broome, R., Corbett, S., & Smith, W. (2012). Emergency department visits, ambulance calls, and mortality associated with an exceptional heat wave in Sydney, Australia, 2011: A time-series analysis. *Environmental Health*, *11*(1), 3. doi:10.1186/1476-069X-11-3 PMID:22273155

Solberg, K. (2010, October). Solberg (2010). Worst floods in living memory leave Pakistan in paralysis. *Lancet*, *376*(9746), 1039–1040. doi:10.1016/S0140-6736(10)61469-9 PMID:20879089

Uejio, C. K., Wilhelmi, O. V., Golden, J. S., Mills, D. M., Gulino, S. P., & Samenow, J. P. (2011). Intra-urban societal vulnerability to extreme heat: The role of heat exposure and the built environment, socioeconomics, and neighborhood stability. *Health & Place*, *17*(2), 498–507. doi:10.1016/j.healthplace.2010.12.005 PMID:21216652

Uejio, C. K., Wilhelmi, O. V., Golden, J. S., Mills, D. M., Gulino, S. P., & Samenow, J. P. (2011). Intra-urban societal vulnerability to extreme heat: The role of heat exposure and the built environment, socioeconomics, and neighborhood stability. *Health & Place*, *17*(2), 498–507. doi:10.1016/j.healthplace.2010.12.005 PMID:21216652

Warraich, H., Zaidi, A. K. M., & Patel, K. (2011). Floods in Pakistan: A public health crisis. *Bulletin of the World Health Organization*, *89*(3), 236–237. doi:10.2471/BLT.10.083386 PMID:21379421

Warwick, R. A. (1975). Drought hazard in the United States: A research assessment: Boulder, Colorado, University of Colorado, Institute of Behavioral Science, Monograph no. NSF/RA/E-75/004.

Wilhite, D. A., & Lantz, M. H. (1985). Understanding the drought phenomenon: The role of definitions. *Water International*, *103*(3), 111–120. doi:10.1080/02508068508686328

WMO (World Meteorological Organization). (1975). *Drought and Agriculture. Report of the CAgM Working Group on the Assessment of Drought prepared by C. E. Hounam Chairman, J. J. Burgos, M. S. Kalik, W. C. Palmer, & J. Rodda. Technical Note No. 138, WMO No. 392*. Geneva, Switzerland: World Meteorological Organization.

Xu, Y., Dadvand, P., Barrera-Gómez, J., Sartini, C., Marí-Dell'Olmo, M., Borrell, C., ... Basagaña, X. (2013). Differences on the effect of heat waves on mortality by sociodemographic and urban landscape characteristics. *Journal of Epidemiology and Community Health*, *67*(6), 519–525. doi:10.1136/jech-2012-201899 PMID:23443960

Zaitchik, B. F., Macalady, A. K., Bonneau, L. R., & Smith, R. B. (2006). Europe's 2003 heat wave: A satellite view of impacts and land–atmosphere feedbacks. *International Journal of Climatology*, *26*(6), 743–769. doi:10.1002/joc.1280

Chapter 2
Maternal Determinants of Childhood Stunting:
The Case of Pakistan

Sarin Ishaque
https://orcid.org/0000-0001-9573-547X
National University of Sciences and Technology, Pakistan

Junaid Ul Mulk
National University of Sciences and Technology, Pakistan

Muhammad Ali
https://orcid.org/0000-0003-0668-5699
National University of Sciences and Technology, Pakistan

Ashfaq Ahmad Shah
National University of Sciences and Technology, Pakistan

ABSTRACT

The progress on reducing stunting is rather slow in Pakistan despite significant reductions in poverty which begs a question why Pakistan has been unable to make significant strides in improving nutrition indicators for children over the past few decades. Despite the recognized importance of the problem in national and international forums, research on determinants of child stunting in Pakistan is scarce, especially in the context of the role of mother's health, education, and empowerment in determining a child's nutrition status. Therefore, this chapter incorporates the mother's health, education, and empowerment-related factors in determining the factors that affect child stunting in Pakistan. Using simple OLS methodology on DHS (2012-13) dataset for Pakistan, the authors' results show that improvement in mother's health, women empowerment, and women's education are likely to reduce stunting. Moreover, better hygiene and better food intake also reduce stunting among children in Pakistan.

DOI: 10.4018/978-1-7998-2197-7.ch002

INTRODUCTION

Under-nutrition refers to a state of not having enough food or consuming food with deficiency of substances necessary for growth and health. It is a major public health problem that is responsible for 45% of all deaths in children who are aged below five years of age. Around two million children under the age of 5 die each year due to undernutrition around the world (García Cruz et al., 2017). According to UNICEF (1991) "stunting reflects the chronic undernutrition during the most critical periods of growth and development in early life". Stunting is defined as having a low height for age that mostly affects children that are low or deprived of micronutrients. It takes place in children under the age of two and has a chance of not being reversed, if not treated and is linked with slow development and impaired cognitive function. Inadequate nutrition over a long period of time or chronic illnesses could lead to stunting. Stunting can be measured in children aged between 0 to 59 months if their height-for-age is below -2 standard deviations of the world health organization (WHO) child growth standards median. This is a repetitive process because women who were themselves stunted in childhood had a higher chance of having stunted children, making this an intergenerational cycle. Research suggests that investing in interventions on access to WASH (water, sanitation, and hygiene) for baby and mother could play a pivotal role in reducing stunting (Mbuya and Humphrey, 2016; Hernández-Díaz et al, 1999). Improving family and community hygiene practices also reduces the likelihood of childhood stunting. If there is proper sanitation then there will be less chance of childhood illness such as diarrhea, which is an important factor causing stunting. There is a need to improve breastfeeding practices which is one of the most important nourishment for a child. Factors such as women's nutritional status, especially for the pregnant mothers and adolescent girls need to improve as studies prove mother nutritional status to have a significant impact on stunting (WHO 2018).

The childhood stunting rate in Pakistan has been relatively persistent over the years and it is among the highest in the region despite the fact that poverty, WASH and health indicators have improved over the same period (Asim and Nawaz, 2018). Keeping this in view, this study aims to identify the determinants of stunting in Pakistan. In particular, the focus of this study is on mother's nutrition, education, empowerment, and health related indicators to ascertain whether stunting is an intergenerational phenomenon. The data used for the econometric model is extracted from the Demographic and Health Survey (DHS) of Pakistan from 2012-2013. DHS is a nationally representative survey that has information regarding the key determinants of stunting that will be used in the estimation of the econometric models in this study.

Figure 1.

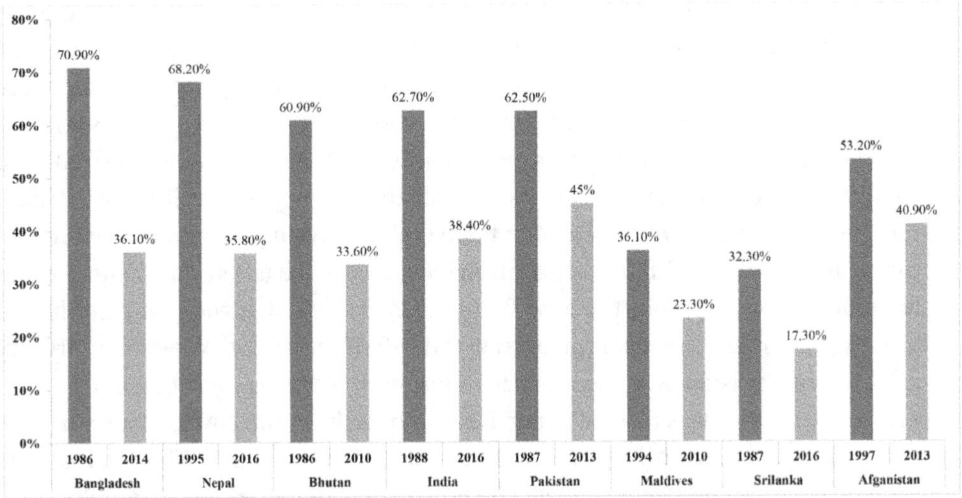

The rest of the chapter is organized as follows. Section 2 provides statistics on the prevalence of under-five stunting in Pakistan relative to other South Asian countries. Section 3 discusses the econometric model, variables and data. Section 4 presents and discusses results and Section 5 concludes.

PREVALENCE OF STUNTING IN PAKISTAN

About 37% of children aged below 5 years residing in South Asia[1] continue to face stunted growth especially among children from rural areas (WHO 2019). Stunting is more common among households, which face multiple types of disadvantages such as low diversification in dietary intake, poor maternal education, and household poverty. According to JME (Joint Child Malnutrition Estimates) data set estimates of malnutrition, 3 out of 8 countries namely Bangladesh, Nepal, and Bhutan have successfully been able to reduce stunting by around 50% with Bangladesh with the most reduction from 70.9% (1986-88) to 36.1% (2014) followed by Nepal from 68.2% (1995) to 35.8% (2016) and then Bhutan 60.9% (1986-88) to 33.6% (2010). Other countries followed up with India 62.7% (1988-90) to 38.4% (2015-16) then Pakistan 62.5% (1985-87) to 45% (2012-13) then Maldives 36.1% (1994) to 23.3% (2010) and lastly Sri Lanka 32.3% (1987) to 17.3% (2016) and Afghanistan 53.2% (1997) to 40.9% (2013).

As shown in Figure 1, Bangladesh and Nepal have the largest reductions with 2.9 and 4.1 percent point per year respectively. India and Pakistan, on the other hand, have

1.3 and 0.6 percent point reductions per year. In Pakistan, about half of the children below the age of five face an increasing concern of stunting. Stunting is found more in male children (48%) than in female children (42%). A 3.3% increase in stunting has been found in Pakistan since 2001 whereas throughout the world malnutrition and stunting stats are decreasing. Overall the lack of public and private spending in both health care and education sectors along with increasing socio-economic inequalities between urban and rural sectors of the country makes it more difficult to fight problems like stunting. Despite Pakistan's improving economic conditions, the rate of maternal and childhood nutrition indicators have fallen way behind others in the region. Pakistan ranks the lowest among countries on a wide range of health indicators particularly neonatal, maternal and infant mortality. Neonatal mortality rate (NMR) in Pakistan is 46 deaths per thousand live births on average. For rural areas of Pakistan NMR is 62 deaths per 1000 live births while NMR for urban areas is 47 deaths per thousand live births. The urban-rural ratio of NMR for Pakistan stands at 0.8 (UNICEF, Maternal and New Born Health Disparities in Pakistan – country profile). Therefore, the poor state of Pakistan's health care system is a major hindrance to stunting reduction in the country (IARAN 2017). Studies have found that neonatal and maternal factors are the early determinants of stunting and neonatal in particular has proven to be a more important factor in determining stunting than prenatal (Mal-Ed Network Investigators, 2017).

Research has shown the mother related factors plays an important role in the upbringing of the child especially in the early stages of the child's life. Factors such as literacy of women, age at marriage and place of delivery can have a direct impact on the child's health and can ultimately lead to stunting. In the DHS report (2012-2013) it was found that literacy in women prevails to be 68.9% (urban) compared to 30.6% (rural). This results in a smaller number of females being employed and females having a lesser say in marriage decisions and family planning. Female education is not given much importance and women are not allowed much freedom in leading life the way they would want. All of these factors take a direct toll on a child's health, leading to stunting.

Figure 2 presents the intergenerational stunting cycle. The green pathway shows the period between conception and 2 years ('the first 1000 days') when preventable measures have the most effect on stunting. The yellow path shows the period between age 2 and mid-childhood when there may be some catch-up in linear growth whereas the shorter yellow path before *Conceptus* shows that improvement in the diet of the stunted women during the pre-conception period positively affects the births. The red pathway shows a period where interventions do not affect stunting. Blue boxes list all factors associated with age while white boxes show outcomes associated with age. The dashed line, between 2 years and adulthood, denotes a stunted child to gain excess weight as a result of abundant access to food. The solid line depicts

Figure 2.

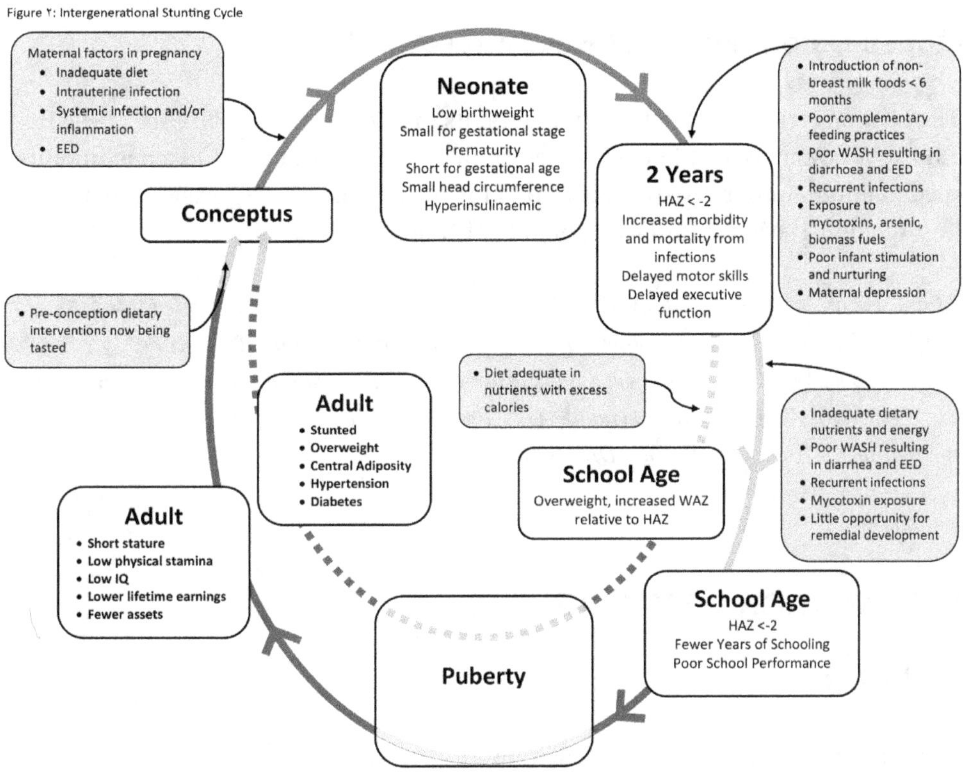

an environment in which stunted child is deprived of food. Mother's undernutrition accounts for about 20% of childhood stunting (Prendergast and Humphrey, 2014). Poor nutritional diet of women before and during pregnancy leads to low birth rates. Women's poor nutrition before and during pregnancy contributes to low birth weight and a high probability of stunting. Evidence reveals that lower educational standards, low decision-making autonomy, early age of marriage, interventions in employments of women and domestic violence harm child growth (WHO 2019). However, it is found that the first 2 years of the child are revealed to be most sensitive to stunting and proper mother's nutritional care is one of the most important factors which ensure that a child's growth is not impaired. Ill health may lead to poor dietary intake causing undernutrition in women which contributes to undernutrition in children ultimately leading to stunting.

ECONOMETRIC MODEL

Model: Determinants of Height-for-Age Z-score

The econometric model in this study took its motivation from UNICEF (1991) as it aims to accommodate all possible determinants of child stunting in a parsimonious specification. Authors used the equation 1 for estimation of the regression coefficients:

Equation 1. Model for Determinants of Stunting:

$$\begin{aligned}
Malnutrition_i &= \beta_0 + \beta_1 Child\,Age_i + \beta_2 No.of\,Under\,5\,Children_i + \beta_3 Child\,is\,Male_i \\
&+ \beta_4 Water\,Treatment_i + \beta_5 Mother's\,Age_i + \beta_6 Mother\,has\,Primary\,Education_i \\
&+ \beta_7 Improved\,Water_i + \beta_8 Improved\,Sanitation_i + \beta_9 Handwashing\,Facility_i + \beta_{10} Rural\,HH_i \\
&+ \beta_{11} B40_i + \beta_{12} Women\,Empowerment_i + \beta_{13} Prenatal\,Care_i + \beta_{14} Immunization\,of\,child_i \\
&+ \beta_{15} Mother's\,Health_i + \beta_{16} Food\,Diversity + \epsilon_t
\end{aligned} \tag{1}$$

EXPLANATION OF VARIABLES

Child Age

This variable shows the age of each child in months taken from a total of 3,975 households from the year 2012-2013.

Number of Under 5 Children

This variable shows the count of children in a household whose age is less than five years.

Child is Male

If the child is male, it is represented by the number 1 and if it is a female, then it is represented by the number 0.

Water Treatment

Water is considered properly treated if it is boiled or chlorinated. Two variables were chosen from the data of DHS to capture water treatment behavior. Water treatment 1 equals 1 if the water is boiled or chlorinated. If any one of the two treatments is done, it is represented by the number 1 while if both were not done then it was represented by the number 0. Water treatment 2 equals 2 if any kind of treatment is done to the water to make it safe for drinking. If the answer is yes, it is represented by the number 1 but if the answer is no then it is represented by the number 0.

Mother's Age

This variable shows the mother's current age in years.

Mother Has Primary Education

Mothers, who had completed primary and above schooling were 52% less likely to have stunted children than mothers with no formal education (Musbah &Worku, 2016). Therefore, if the mother has no education, it is represented by the number 0 and if she has primary or higher education then it is represented by the number 1.

Improved Water

Improved drinking water sources are those which, by definition, have the potential to deliver safe water JMP Joint Monitoring Program. This JMP definition is used as a benchmark to compare the different levels of services across the countries. Lack of clean and safe water and sanitation facilities are likely to create diarrhea in children (Prüss-Ustün et al. 2014).

Unimproved water causes different diseases such as diarrhea and hence leads to stunting. If the child is provided with contaminated/unimproved water, then there is a high chance of the child getting stunted. From the given list of water sources provided in the DHS report, the water sources are classified into two types; improved and unimproved. Improved water includes; protected spring, piped into dwelling, piped to yard/plot, public tap, tube well or borehole, hand pump, protected well, rainwater and filtration plant. The list of unimproved water includes river/dam/lake, tanker truck, and cart with a small tank, bottled water, unprotected well and unprotected spring. The above classification is according to the JMP definition. Improved sources of water are represented by the number 1 while unimproved sources of water are represented by the number 0.

Improved Sanitation

According to the JMP 2017 report, sanitation services refer to the management of excreta from the facilities used by individuals, through emptying and transport of excreta for treatment and eventual discharge or reuse. The lack of sanitation may enhance disease transmission by contaminating water sources, increased person to person transmission via hands/food and flies (Cairncross, 2003; Ezzati et al. 2005). Improved sanitation facilities are those designed to hygienically separate excreta from human contact. There are three main ways to meet the criteria for having a safely managed sanitation service (WHO). People should use improved sanitation facilities which are not shared with other households, and the excreta produced should either be:

- treated and disposed of in situ,
- stored temporarily and then emptied and transported to treatment off-site, or
- transported through a sewer with wastewater and then treated off-site.

The toilet types are categorized as "improved" and "unimproved sanitation" using the toilet definitions provided in the DHS survey. Improved sanitation includes flush to a piped sewer system, flush to a septic tank, flush to pit latrine, ventilated improved pit latrine, pit latrine with slab and flush toilet. List of unimproved sanitation includes flush to somewhere else, flush-don't know where, pit latrine without slab-open latrine, no facility-bush/field, bucket toilet, hanging toilet and other. Improved sanitation is represented by the number 1 while the unimproved sanitation is represented as 0.

Hand Washing Facility (Hygiene)

The presence of a handwashing facility with soap and water on-premises has been identified as the priority indicator for global monitoring of hygiene. Households that have a handwashing facility with soap and water available on-premises will meet the criteria for a **basic** hygiene facility. Households that have a facility but lack water or soap will be classified as having a **limited** facility and distinguished from households that have no facility at all (JMP). The handwashing facility variable is a composite index of three variables; a place where they wash their hands, the presence of water and the presence of soap or detergent. If all three indicators are present, then the handwashing variable takes the value 1 and 0 otherwise.

Rural Household

Broeque (2007) has studied the differences between the rural and urban areas of Mozambique. His study concludes that since rural areas are deprived of many resources and also lack in education, they are likely to have a high rate of stunting as compared to the urban areas. The variable "type of place of residence" in the DHS data had two types; urban and rural. The variable for rural areas is represented by the number 1 when the household is in rural area and 0 otherwise.

B40 (Bottom 40 Wealth)

The wealth/asset index is divided into 5 quintiles. The population in the first 20% are the richest, the second 20% are the rich, the third 20% are the middle while the fourth 20% of the group are poor and the last 20% are the poorest. B40 is the bottom 40 or the last 2 quintiles of wealth. If a household falls into the category of either poor or poorest, it is represented by the number 1. While if it falls in any of the remaining 3 categories (middle, rich and richest), it is then represented by the number 0.

Women Empowerment

Women empowerment has a major role to play in child stunting. A woman who has the deciding power is more likely to make decisions in favor of her child and her health. Three variables are used in this study to capture women empowerment. The first variable chosen is "person who usually decides how to spend the respondent's earnings". If the decision is taken by the respondent alone or the respondent and husband/partner, then it is represented by the number 1. But if the decision is taken by the husband/partner alone or someone else or family elders then it is represented by the number 0.

The second variable chosen is "person who usually decides on the respondent's health care". The reason behind choosing this variable is that higher confidence of the mothers in easily accessing the health care services led to fewer chances of having a stunted child (Kamiya et al., 2018). If the decision is taken by the respondent alone or the respondent and husband/partner, then it is represented by the number 1. But if the decision is taken by the husband/partner alone or someone else or family elders then it was represented by the number 0.

The third variable is "Decision maker for using contraception" if the decision was taken by the respondent alone or the respondent and husband/partner then it was represented by the number 1. But if the decision was taken by the husband/partner alone or someone else or family elders then it was represented by the number 0.

Prenatal Care

The WHO recommends at least four prenatal visits to the hospital by a pregnant woman. From the DHS data if the woman had made 0 to 3 visits to the doctor it is represented by the number 0 while if she made 4 to 60 visits then it is represented by the number 1.

Immunization of Child

Child immunization helps to protect the child from various diseases. From the DHS data if the child ever had a vaccination it is represented by the number 1 but if it never had a vaccination then it is represented by the number 0.

Mother's Height

This variable represents mother's height measured in centimeters.

Malnutrition

This study uses three dependent variables: 1) Height-for-Age Z-scores, 2) moderate and severe stunting (HAZ <-2SD) and 3) severe stunting (HAZ <-3SD). It is important to note that the interpretation of coefficients is different for HAZ variable where the increase in HAZ is a positive development while for stunting variables whereas, movement from 0 to 1 is a negative development. Since the dichotomous characterization of HAZ values results in loss of variance, we chose to use continuous as well as dichotomous variables to test the robustness of the results.

Mother's Health

Mother's health which is one of the most important factors in determining child stunting and one which directly affects it is captured by creating two separate variables. The first variable is "women stunted". According to Hernåndez-Díaz et al., (1999) women who have short stature have a higher probability of having low birth weight (LBW) infants. LBW, in turn, are at risk for poor growth during childhood. For this variable, the height (cm) of the respondents (women) were divided by 10 and any height below 145 cm was represented by 1 (categorized as stunted) and heights above 145cm was represented by 0 (categorized as not stunted).

The second variable created under mother's health is "mom underweight". Women who have a poor dietary intake, are underweight, stunted or are not able to gain a healthy amount of weight during their pregnancy are more likely to have children with restricted growth (Duggan, 2003). BMI scores of the respondents (mothers) were divided by 100 and any BMI below 18.5 was represented by 1 (categorized as underweight) and BMI above 18.5 was represented by 0 (categorized as not underweight).

Food Groups

"Food adequacy is a concept that includes three distinct components: (1) that all nutritional needs should be satisfied, both in terms of energy and all other essential nutrients; (2) that foods consumed should be both safe and palatable, and; (3) that the types of foodstuffs commonly available should be acceptable within the prevailing food culture" (Oshaug, (1994) as cited by Rue et al., (1998)).

The WHO (2008) definition of diet diversity of food groups (for children under 24 months) is based on the consumption of seven food groups consumed during the past 24 hours. The seven food groups considered are (1) grains, roots and tubers; (2) legumes and nuts; (3) dairy products; (4) flesh foods including organ meats; (5) eggs; (6) Vitamin A rich fruits, and vegetables including orange and yellow vegetables; and (7) and other fruits. To have a minimum acceptable diet, a child needs to have consumed from at least four of the seven food categories.

Each food group is then made into a binary form. Each group, if consumed, by the child, is represented by the number 1 whereas if it is not consumed, then it is represented by the number 0. Furthermore, a variable labeled as "food diversity" is created in which if 4 or greater than 4 groups out of 7 are consumed it is represented by the number 1 while if less than 4 groups are consumed it is represented by the number 0.

Data

The Demographic and Health Survey program collects and publishes nationally representative household-level data for a wide range of monitoring and impact evaluation indicators in the areas of population, health, and nutrition in developing countries. The 2012-13 Pakistan Demographic and Health Survey is the third survey conducted as part of the DHS international series. The target population of this study is the population of Pakistan excluding Azad Jammu and Kashmir, FATA, and restricted military and protected areas based on the information of the year 2012-2013 only. It consists of urban and rural areas of the four provinces of Pakistan and includes Gilgit Baltistan as defined in the 1998 Population Census.

Table 1 provides descriptive analysis of the variables used in this study. All of the variables have been converted into binary terms with the maximum value of 1 and the minimum of 0 except 4 variables which include: HAZ (height for age) Mothers age (momage), Age of child in months (childage) and children under the age of 5 (under 5 children).

Children with HAZ scores below or close to -2 are considered to be moderately and severely stunted while HAZ scores below -3 portray severe stunting. The maximum score of HAZ is seen to be 5.81 while the minimum score is seen to

Table 1. Descriptive Statistics

Variable	Obs	Mean	Std. Dev.	Min	Max
HAZ	3,071	-1.8	1.9	-6	5.81
Moderate and Severe	3,071	0.4	0.5	0	1
Severe	3,071	0.3	0.4	0	1
Mother's Education	18,186	0.6	0.5	0	1
Women Empowerment	11,607	0.6	0.5	0	1
Prenatal Care	7,443	0.4	0.5	0	1
Women Stunted	4,140	0.1	0.2	0	1
Mom Underweight	4,117	0.1	0.3	0	1
Mom Age	4,272	29.4	6.1	16	49
Boiled or Chlorinated	18,158	0.1	0.3	0	1
Any Treatment	18,159	0.1	0.3	0	1
Improved Water	18,186	0.9	0.3	0	1
Improved sanitation	11,331	0.7	0.4	0	1
Hand wash	18,186	0.5	0.5	0	1
Food Diversity	7,121	0.1	0.3	0	1
Immunization	8,130	0.9	0.4	0	1
Rural	18,186	0.5	0.5	0	1
Wealth Indicator	11,763	0.4	0.5	0	1
Child Age	3,957	29.9	17.3	0	59
Under 5 Children	11,763	2.4	1.5	0	13
Male	11,763	0.5	0.5	0	1

be -6. The average HAZ score is calculated to be -1.78 of the sample set which indicates that the majority of the children in the sample set suffer from moderate and severe stunting.

The current age of the youngest respondent (mother) from the DHS report is reported to be 16 while the maximum age of the respondent (mother) is reported to be 49 from the table above. The mean age of mothers in the sample is found to be 29 years. Child age in the DHS report has been measured in months with the minimum value of 0 and the maximum value of 59 months. The mean age of children in the sample is found to be 30 months. The maximum number of children under the age of 5 in one household is seen to be 13 while the minimum is 0 whereas the average number of children under the age of 5 in one household is 2.

Table 2. Reports the pairwise correlation coefficient between variables

	1	2	3	4	5	6	7	8	9	10	11	12	13	14	15	16	17	18	19	20	21
1.HAZ	1																				
2. Moderate and severe	-0.8	1.0																			
3. Severe	-0.7	0.7	1.0																		
4. Mother's Education	0.1	-0.2	-0.1	1.0																	
5. Women empowerment	0.1	-0.1	-0.2	0.1	1.0																
6. Prenatal Care	0.1	-0.1	-0.1	0.4	0.2	1.0															
7. Women Stunted	-0.1	0.1	0.0	0.0	0.0	0.0	1.0														
8. Mom Underweight	-0.1	0.1	0.0	-0.1	0.0	0.1	0.0	1.0													
9. Mom Age	0.0	0.0	0.0	-0.1	0.0	-0.1	-0.1	-0.1	1.0												
10. Boiled or Chlorinated	0.1	-0.1	-0.1	0.2	0.1	0.2	0.0	-0.1	0.0	1.0											
11. Any Treatment	0.1	-0.1	-0.1	0.2	0.0	0.1	-0.1	-0.1	0.0	0.8	1.0										
12. Improved Water	0.0	-0.1	-0.1	0.1	0.1	0.1	0.0	0.1	0.0	0.0	0.0	1.0									
13. Improved Sanitation	0.2	-0.1	-0.1	0.3	0.0	0.2	0.0	-0.1	0.0	0.1	0.1	0.2	1.0								
14. Handwash	0.2	-0.2	-0.2	0.4	0.1	0.2	0.0	-0.1	0.0	0.2	0.1	0.2	0.3	1							
15. Food Diversity	0.1	0.0	0.0	0.1	0.1	0.1	0.0	-0.1	0.1	0.0	0.0	0.0	0.1	0	1						
16. Immunization	0.0	0.0	-0.1	0.1	0.2	0.1	0.0	0.1	0.0	0.1	0.0	0.1	0.1	0.1	0.1	1					
17. Rural	-0.1	0.1	0.1	-0.3	-0.1	-0.2	0.1	0.1	0.0	-0.2	-0.3	-0.1	-0.3	-0.3	-0.1	-0.1	1				
18. Wealth Indicator	-0.2	0.2	0.2	-0.5	-0.1	-0.3	0.1	0.1	0.0	-0.2	-0.1	-0.2	-0.4	-0.5	-0	-0.1	0.4	1			
19. Child Age	-0.2	0.2	0.1	0.0	0.1	-0.1	-0.1	0.0	0.1	0.0	0.0	0.0	0.0	0	0.3	0.3	-0.1	-0.1	1		
20.Under 5 Children	-0.1	0.0	0.0	-0.1	-0.1	-0.1	0.0	-0.1	0.1	-0.1	0.0	-0.1	0.0	-0	0	-0.1	-0	-0.1	-0	1	
21. Male	0.0	0.1	0.0	0.0	0.0	0.0	0.0	0.0	0.0	0.1	0.0	0.0	0.0	0	-0	0	-0.1	-0	0	-0.1	1

The correlation table in Table 2 presents pairwise correlation coefficients between all explanatory variables. Since all of the correlation coefficients are less than 0.6, we conclude that multicollinearity does not exist in our models.

RESULTS AND DISCUSSION

The results in Table 3 are divided into 6 Models. Values of Height for Age (Models 1 and 4) are calculated as continues nonbinary numbers while Moderate and Severe and Severe are calculated in binary numbers with minimum value 0 and maximum value 1. Since information on food consumption of the children is limited to index children under 2 years of age, inclusion of food groups in the regression significantly reduces the sample size. Therefore, in Model 1, 2 and 3, the dependent variables; Height for Age (HAZ), Moderate and Severe Stunting and Severe Stunting, are regressed on the determinants of stunting including the food diversity variable and in Model 4, 5 and 6 excluding the food diversity variable to allow for a robustness check of the results.

We found partial support for the claim that malnourished women tend to have malnourished children. In models 4-6 without food groups and vaccinations, we show that children of stunted or underweight women are likely to be stunted. This result, however, does not hold for the smaller sample size when food groups and vaccinations are included in the models (1-3). Our results could have two possible implications; first, it is possible that the insignificant relationship between mother's and children's nutrition is due to the reduction in sample size and second, perhaps the insignificance of the two variables for mother's nutrition is due to omitted variable bias. Since adequate food consumption is vital for child growth, we show that number of food groups consumed has a negative relationship with moderate and severe stunting and positive relationship with HAZ. Therefore, we believe that the reason behind the insignificance of the mother's characteristics is omitted variable bias. It should be noted that, from DHS dataset, it was impossible to know whether the stunted mothers were genetically born stunted or not therefore our findings are valid for one of the most important non-genetic factors that effects stunting is adequate nutritional intake. Without proper nutritional intake, a mother may grow up to be stunted. This will have adverse growth effects on the growth of their child resulting in an intergenerational stunting phenomenon.

Our results show partial support for the claim that children of the educated mothers are likely to be taller than their counterparts. For a larger sample of children in Models 4 and 5, we found that educated mothers are likely to have taller children. These results, however, are not robust across models. Our results also show that children of the women who are empowered in decision making are less likely to be stunted. The result is relatively robust across specifications with only Model 2 showing an insignificant relationship between women empowerment and moderate and severe stunting. Our results show that the lack of women empowerment is one of the key determinants of childhood stunting in Pakistan. Among the WASH indicators, only handwashing has a significant relationship with child's height i.e. children living in the households with a handwashing facility with soap and water are less likely to be stunted. Water treatment, access to improved water, and improved sanitation did not have an impact on child stunting in Pakistan.

Table 3. Regression Results using OLS methodology

	With Food Groups			Without Food Groups		
	HAZ (1)	Moderate and Severe Stunting (2)	Severe Stunting (3)	HAZ (4)	Moderate and Severe Stunting (5)	Severe Stunting (6)
Mother's Education	-0.424	0.115	0.0691	0.268*	-0.0959**	-0.0633
	(-1.77)	(1.88)	(1.22)	(2.22)	(-2.59)	(-1.89)
Women Empowerment	0.425*	-0.0901	-0.172***	0.377***	-0.0705*	-0.118***
	(2.48)	(-1.84)	(-3.56)	(3.64)	(-2.17)	(-3.91)
Prenatal Care	0.535*	-0.0872	-0.0545	0.353**	-0.0578	-0.0489
	(2.40)	(-1.43)	(-1.10)	(2.92)	(-1.69)	(-1.73)
Women Stunted	-0.510	0.145	-0.0304	-0.548**	0.127*	0.0904
	(-1.88)	(1.31)	(-0.33)	(-3.04)	(2.04)	(1.51)
Mom Underweight	-0.245	0.0844	-0.0285	-0.429***	0.0907*	0.0542
	(-1.47)	(1.33)	(-0.49)	(-3.75)	(2.02)	(1.34)
Mom Age	0.00141	0.000966	0.00194	0.00924	0.000352	0.000204
	(0.10)	(0.21)	(0.54)	(1.05)	(0.13)	(0.10)
Boiled or Chlorinated	0.604	-0.178	-0.0372	-0.279	0.0891	0.168
	(0.43)	(-0.70)	(-0.15)	(-0.58)	(0.49)	(1.85)
Any Treatment	-0.445	0.161	0.0414	0.110	-0.106	-0.0917
	(-0.32)	(0.63)	(0.18)	(0.25)	(-0.62)	(-1.19)
Improved Water	0.0727	0.0655	-0.0473	0.115	0.00782	-0.0555
	(0.24)	(0.87)	(-0.67)	(0.60)	(0.16)	(-1.20)
Improved Sanitation	0.0121	0.0439	0.0367	-0.0488	0.0327	0.0392
	(0.06)	(0.71)	(0.68)	(-0.41)	(0.85)	(1.25)
Hygiene	0.671**	-0.192***	-0.0997*	0.264**	-0.0862**	-0.0800***
	(3.33)	(-3.65)	(-2.06)	(2.61)	(-2.60)	(-3.33)
Food Diversity	0.849**	-0.163*	-0.0979			
	(2.92)	(-2.16)	(-1.46)			
Immunization	0.343	-0.0450	-0.127*			
	(1.50)	(-0.60)	(-2.00)			
Rural	-0.223	0.00941	0.0579	0.0815	-0.0587	-0.0136
	(-0.97)	(0.14)	(1.04)	(0.70)	(-1.76)	(-0.45)
Wealth Indicator	-0.217	0.165*	0.0697	-0.319*	0.138**	0.0831*
	(-0.83)	(2.25)	(1.14)	(-2.30)	(3.08)	(2.06)
Child Age	-0.0727***	0.0161***	0.0110**	-0.0228***	0.00400***	0.00254**
	(-4.86)	(3.78)	(3.06)	(-6.60)	(3.86)	(3.30)
Under 5 Children	0.0704	-0.00193	-0.0180	-0.0299	0.00432	0.00263
	(0.91)	(-0.10)	(-1.13)	(-0.80)	(0.39)	(0.27)
Male	-0.253	0.101*	0.0243	-0.281**	0.0817**	0.0451
	(-1.53)	(2.03)	(0.57)	(-2.79)	(2.77)	(1.62)
Constant	-1.643*	0.183	0.311	-1.763***	0.374***	0.293**
	(-2.21)	(0.94)	(1.64)	(-4.33)	(3.34)	(2.69)
N	540	540	540	1830	1830	1830

t statistics in parenthesis

*p<0.05 **p<0.01 ***p<0.001

Source: (As calculated by Authors)

Differentiating between genders, our results show that a male child is 0.28 standard deviation shorter as compared to a female child. One of the reasons behind this may be because due to the preference of a male child, they are more taken care of and are more privileged and overfed to keep them healthy and nourished. However, overfeeding causes diarrhea and as a result, a child continuously loses the essential nutrients required for growth. This along with other factors may cause stunting to prevail more in males than in females.

CONCLUSION

About half of the children under the age of 5 in Pakistan are exposed to the increasing concern of stunting. Despite the reduction in poverty in Pakistan in recent years, there has been little or no progress towards the reduction in stunting. This study was conducted to analyze the intergenerational phenomenon of a stunting cycle in Pakistan by incorporating the mother's nutrition related indicators along with other determinants of stunting. The gap was addressed by using a simple regression framework, inspired by UNICEF (1990) framework, on data from DHS (2012-13). The results of the analysis show that improvements in women's health and prenatal care are likely to bring a significant reduction in child stunting. Both these variables have a negative impact on stunting implying that a health woman is less likely to have a stunted child. Moreover, the findings of this study show that a mother's education plays a vital role in reducing child stunting. The results also show that stunted mothers are likely to have a stunted offspring as well as showing a need for immediate action to address child stunting today to prevent stunting in future generations. Among the WASH indicators, adequate hygiene is the only significant determinants of stunting. The overall conclusion of this study is that mother's health, education and empowerment are the key determinants of childhood stunting and interventions should target education and health of mothers to reduce the rate of childhood stunting in Pakistan.

While this study provides strong results and recommendations, there were some limitations as well. First, as cross-sectional data was used for estimations, the results of this study should be interpreted as correlations instead of causations. Second, inclusion of food and vaccinations in the model significantly reduced the sample size which reduces the comparability of the two sets of models presented in the regression table. Lastly, the DHS data does not have information on genetic stunting of children and mother therefore pre-birth likelihood of being stunted cannot be ascertained.

FUTURE RESEARCH DIRECTIONS

Considering the incomes of the individuals, the results showcase that increase in the wealth of a person directly affects child stunting by reducing it. An important finding of the study is that with the inclusion of wealth indicator, the rural dummy loses its significance showing the rural-urban differences are likely to be driven by income disparities.

The results of this study have a number of policy implications. First, there is a need to take measures to improve maternal health care in the country. Second, enrollment rates for females should be increased as educated mothers are likely to take better care of their children. Third, gender equality has to be advocated because children of empowered women are less likely to be stunted. Among WASH indicators, it is shown that better hygiene is a strong determinant of child height. Access to improved water and improved sanitation did not have significant relationship with child height hence use all three proxies of malnutrition.

REFERENCES

Cairncross, S., Shordt, K., Zacharia, S., & Govindan, B. K. (2005). What causes sustainable changes in hygiene behaviour? A cross-sectional study from Kerala, India. *Social science & medicine (1982)*, *61*(10), 2212–2220.

Duggan, M. B. (2003). Nutritional update: Relevance to maternal and child health in East Africa. *African Health Sciences*, *3*(3), 136–143. PMID:14676720

García Cruz, L. M., Gonzalez Azpeitia, G., Reyes Suarez, D., Santana Rodríguez, A., Loro Ferrer, J. F., & Serra-Majem, L. (2017). Factors associated with stunting among children aged 0 to 59 months from the central region of Mozambique. *Nutrients*, *9*(5), 491. doi:10.3390/nu9050491 PMID:28498315

Hernández-Díaz, S., Peterson, K. E., Dixit, S., Hernández, B., Parra, S., Barquera, S., ... Rivera, J. A. (1999). Association of maternal short stature with stunting in Mexican children: Common genes vs. common environment. *European Journal of Clinical Nutrition*, *53*(12), 938–945. doi:10.1038j.ejcn.1600876 PMID:10602351

Kamiya, Y., Nomura, M., Ogino, H., Yoshikawa, K., Siengsounthone, L., & Xangsayarath, P. (2018). Mothers' autonomy and childhood stunting: Evidence from semi-urban communities in Lao PDR. *BMC Women's Health*, *18*(1), 70. doi:10.118612905-018-0567-3 PMID:29789003

Mbuya, M. N. N., & Humphrey, J. H. (2016). Preventing environmental enteric dysfunction through improved water, sanitation and hygiene: an opportunity for stunting reduction in developing countries. *Maternal & Child Nutrition, 12*(Suppl Suppl 1), 106–120.

Musbah, E., & Worku, A. (2016). Influence of maternal education on child stunting in SNNPR, Ethiopia. *Central African Journal of Public Health*, *2*(2), 71.

Prüss, A., Kay, D., Fewtrell, L., & Bartram, J. (2002). Estimating the burden of disease from water, sanitation, and hygiene at a global level. *Environmental Health Perspectives, 110*(5), 537–542. PMID:12003760

KEY TERMS AND DEFINITIONS

Hygiene: Conditions contributing to staying healthy and prevent diseases especially by taking measures for cleanliness.

Intergenerational: Relating to or extending over several generations.

Malnutrition: The lack of proper nutrition either because of lack of food, lack of eating the right nutrients or being unable to use the nutrients that one does eat.

Neonatal death: Death of infants during the first 28 days of life (0-27 days).

Open defecation: The practice of people going out in open spaces to defecate rather than going to toilets.

Socioeconomic factors: Economic factors as dictated by societal structure i.e. education, income, occupation.

Stunting: Under nutrition in the most important years of life; when growth and development is taking place.

WASH indicators: Water, sanitation and hygiene indicators.

Women empowerment: A social action that allows women to take the decisions regarding her life and her children's lives.

ENDNOTE

1 The eight countries in South Asia which include: Pakistan, Nepal, Sri Lanka, Maldives, India, Bhutan, Bangladesh, and Afghanistan

Chapter 3
Climate Change and Diseases of Plants and Animals:
A Study in Home Gardens of West Bengal, India

Sebak Kumar Jana
https://orcid.org/0000-0002-3532-4350
Vidyasagar University, India

Joyashree Roy
https://orcid.org/0000-0002-9270-8860
Asian Institute of Technology, Thailand & Jadavpur University, India

ABSTRACT

Home Garden is a complex sustainable land use system that combines multiple farming components of the homestead and provides environmental services, household needs, and employment and income generation opportunities to the households. Predicted climate changes have serious implications for crop and livestock yields particularly in tropical regions. Home garden may act as a cushion to the adverse climate shocks. There is dearth of in-depth study of home garden ecosystem in India. Authors selected 100 households in Garhbeta-1 block, which is in the dry zone in the district of Paschim Medinipur in West Bengal for this study. The main objectives of this chapter include: (a) identifying key characteristics of the home garden, (b) identifying the pattern of climate change from the household perceptions and the changes made in the home gardens. (c) identifying different diseases of plants and animals in home gardens in the study area (d) managing diseases of plants in home gardens, (e) identifying different problems/ constraints in home.

DOI: 10.4018/978-1-7998-2197-7.ch003

INTRODUCTION

"Home Garden (HG)" is an intricately managed traditional land use system that combines multiple farming components of the homestead primarily practiced to meet diverse household needs, food and nutrition security, employment and income generation but scientifically provide environmental services as co-benefits. Home garden generally is a kind of mixed cropping of fruits, vegetables and trees (Marambe et al 2012, Jana et al 2015). The importance of Home Garden in livelihood generation and in providing food and nutrition security has been discussed by various authors (Crowell *et al.*, 1991, Attygalle 1996, Hitanayake 1996; Label et al 2010, Weerahewa *et al.*, 2011; Pushpakumara*et al.*, 2012, De Zoysa 2014, Jana and Roy 2020). With the efficient utilisation of natural resources like soil, water and sunlight home gardens, can play an important role in providing food and nutrition security to the people. Climate changes if not adapted proactively can have serious implications for crop and livestock yields particularly in tropical regions. It is important to see how home gardens' resilience can act as cushion to the adverse climate shocks.

At current level of climate variability, plant diseases are perceived as problems in home gardens, especially during warm and wet weather seasons. Though most trees, plants, vegetables are susceptible to at least a few diseases, but some diseases are found more susceptible to others. Root rot, leaf spots and fruit rots are the more common diseases for vegetables. Plant diseases may be caused by fungi bacteria, viruses and nematodes. These pathogens are not harmful to humans, but they can be very damaging to plants. Diseases occur when environmental conditions are suitable for pathogens to develop on susceptible plants. Some pathogens attack a wide variety of plants and others attack only specific plants. Also, some pathogens can attack all plant parts, while other attack only selected plant parts. In this backdrop we planned to undertake a study on current status of the home gardens to understand the ecosystem services they provide besides social and economic benefits. Additionally we tried to elicit responses from the home gardeners about threats of pests and diseases.

OBJECTIVES OF THE STUDY

There is very limited number of systematic studies in India on home garden ecosystem and the pattern of diseases found in home gardens. The major objective of the study is to understand the key characteristics of the home garden and diseases of plants and animals in the context of climate variability and change. The objectives of the study include:

Figure 1. Study Area

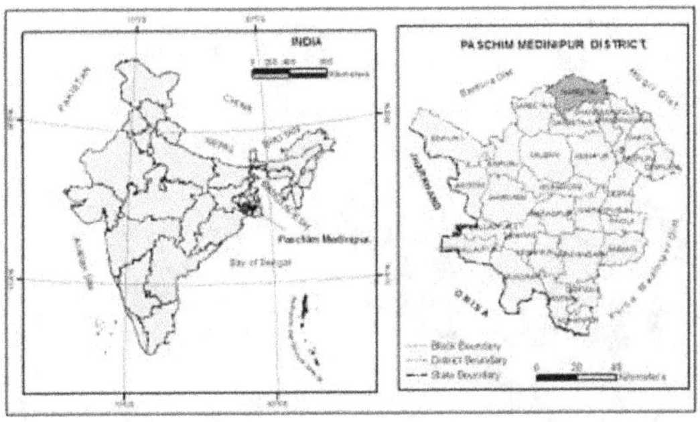

1. Types of plants and animals found currently in the home gardens
2. Perceptions of the home gardeners about climate variability and change
3. Identification of different diseases of plants in home gardens
4. Identification of different diseases of animals in home gardens
5. Management of diseases of plants and animals in home gardens

STUDY AREA

Two villages called Ledagamar village and Keshia village have been selected in Garhbeta – I Block which is situated in the district of Paschim Medinipur in the

Figure 2. Home Garden: Garhbeta, West Bengal, India

Figure 3. Home Garden: Garhbeta, West Bengal, India

state of West Bengal in India. The total number of households surveyed is 100 out of which 74 households were from Ledagamar and 26 are from Keshia. The altitude of the place is 66 meter above the sea level. The soil is sandy loam type. The region is in the dry zone. The summer is hot in these regions (Table 1) . About 36% households live below the poverty line in the block (BDO Office, Garhbeta). In Figure 1, study area has been shown. In Figure 2 and Figure 3, we have presented typical home gardens in the study area.

RESEARCH FRAMEWORK AND METHODOLOGY

The primary survey was carried out in the study area. We had structured questionnaire for our survey. For noting the changes in the home garden we carried out survey in two different seasons. The questionnaire included different aspects relating to home garden characteristics, climate change issues, types of diseases of plants and animals, land use planning, home garden management. The responses were taken from the heads of households. To understand the key characteristics of home gardens and diseases in HGs, we have calculated frequencies, percentages for different parameters of the home gardens.

Figure 4. Trend of minimum temperature in Midnapore
Source: Based on Indian Metriological Department (IMD) data

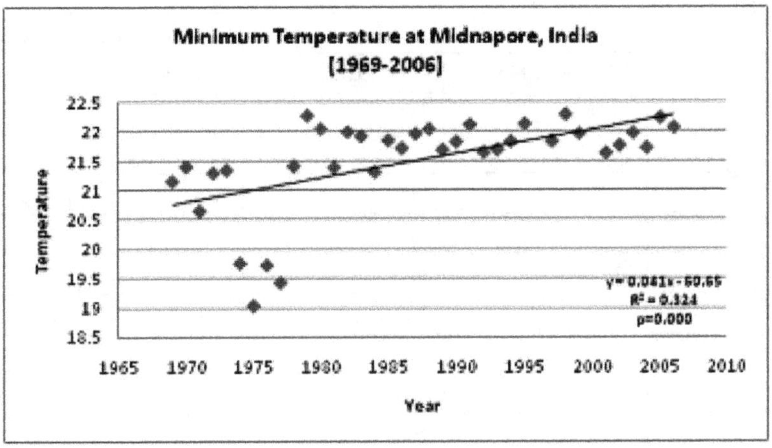

CLIMATE CHANGE AND PROBLEMS OF HOME GARDENS

We have presented in Figure 4 and Figure 5, the trend of change of temperature in Midnapore in West Bengal. It is evident from Figure 4 that minimum temperature during 1969-2006 in the study area (Midnapore) is rising. We have compared month-wise average daily maximum temperature and minimum temperature between two Periods 1969-1990 & 1991-2006 in Table 1. In Figure 6 and Figure 7, we have presented the trends of rainfall in Midnapore.

Figure 5. Trend of maximum temperature at Midnapore
Source: Based on IMD data

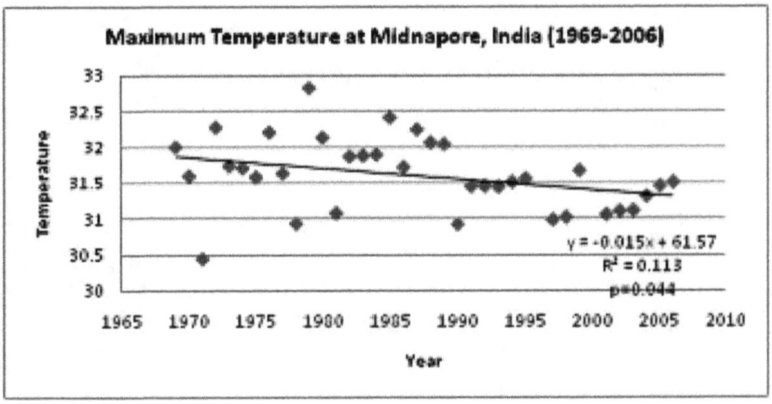

Table 1. Month wise Comparison Average Daily Maximum Temperature between two Periods 1969-1990 & 1991-2006 for Midnapore

Month	Maximum Temperature		Minimum Temperature	
	1969-1990	1991-2006	1969-1990	1991-2006
January	26.1	24.7	13.2	13.4
February	29.1	29.2	16.3	17.3
March	34.2	33.0	20.7	21.6
April	37.6	36.5	24.1	24.7
May	37.1	36.3	25.2	25.9
June	34.8	34.3	25.8	26.4
July	32.3	32.5	25.4	26.2
August	31.8	31.9	25.4	26.1
September	32.0	31.9	25.0	25.7
October	31.7	31.3	22.8	23.6
November	29.4	28.8	18.1	18.5
December	26.3	25.7	13.5	14.1

Source: Estimation based on IMD data

The primary survey in the study area reveals that almost all the households have perceived the change in temperature and rainfall pattern. Perception of households about the different parameters of climate variability and change is presented in Table

Figure 6. Trend of rainfall at Midnapore
Source: Based on IMD data

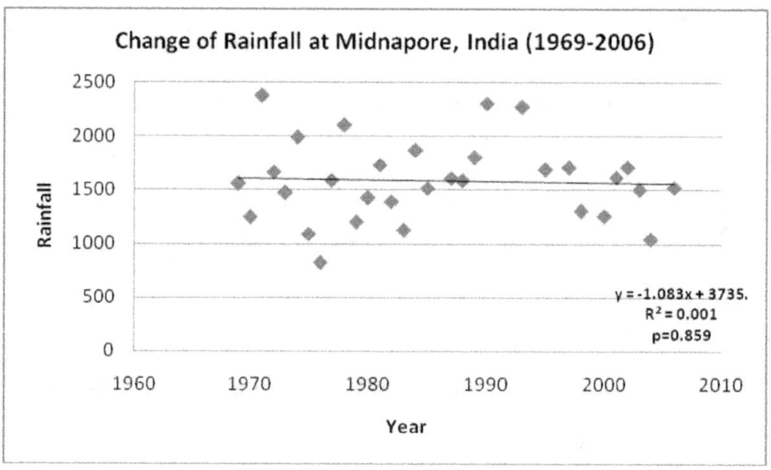

Figure 7. Average rainfall at different weeks in Midnapore
Source: Based on IMD data

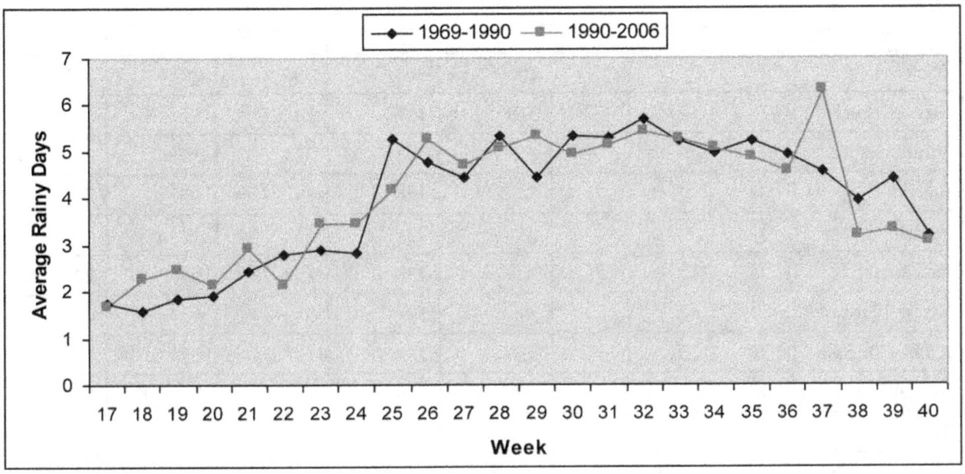

2. We asked the question if they have observed change over a period of 20 years. Almost all households have said that amount of rainfall, period of rainy season, morning temperature, wind turn, mist and dew have decreased over years. On the other hand evening temperature and dry seasons have increased.

In Table 3, we have presented different poultry diseases in Garhbeta region during the period of 2005-06 to 2009-10.During the period, number of diseases of

Table 2. Household perception about climate change

Parameters	Change Observed Over a Period of 20 Years	
	More	Less
Amount of Rainfall	0	100
Rainy Period	0	100
Temperature(Morning)	0	100
Temperature(Evening)	100	0
Wind turn	7	93
Mist	2	98
Dew	2	98
Fog	45	55
Droughts	99	1
Storms	36	64

Source: Primary survey

Table 3. No. of cases detected of different diseases in Regional Poultry Lab in Garhbeta, West Bengal

Aetological Classification	2005-06	2006-07	2007-08	2008-09	2009-10
Bacterial	1019	904	700	457	787
Mycoplasmal	318	137	206	33	102
Viral	513	542	724	288	214
Fungal	74	144	79	39	20
Parasitic	94	92	51	23	81
Nutritional	303	333	178	399	221
Skeletal Disorder	39	52	5	4	0
Kidney Disease	313	237	117	49	29
Miscellaneous Disease	301	226	192	112	100
Total	2974	2667	2252	1404	1554

Source: Department of Animal Husbandry, Government of West Bengal

the poultry in the area has decreased due to additional initiatives taken on part of the Animal Husbandry Department.

SOME CHARACTERISTICS OF HOME GARDENS

The average area of the home gardens under study is 16.8 decimal. Crops, plants and trees occupy the major portion of the home garden area. Average 17% of land is used for different purposes other than home garden including thrashing of crops, playground for the children and other social ceremonial purposes. The number of woody trees per home garden is calculated as 7.17 with maximum of 16 and minimum of 2. Average number of species of non woody trees per home garden is estimated as 1.42. The most frequently found non woody tree in home garden is banana with 58% of households having banana tree.

PESTS AND DISEASES IN PLANTS IN HOME GARDENS

Home garden owners have reported various types of pest attacks, insects and diseases in the home gardens. These are affecting the growth of plants and animals in the home gardens. The most common attack on the home garden is *ledapoka* (77% of households) affecting the vegetables and trees. The other insect and pests are *jab poka, halud poka, sada machhi* etc. The crops in the home garden are vulnerable

Table 4. Number of HHs reporting the Attack of Insects and Pests on Crops and Trees

Name of the Pest/Insect(Local)	English Name	No. of HHs
Leda poka	Semilooper	77
Jab poka	Aphids	34
Haludpoka	-	15
Sadamachi	White Fly	13
Sosakpoka	Plant Hopper	9
Machipoka	Thrips	7
Maker	Mites	6
Makarhsa	-	6
Noli	Shoot Borer	5
Suo poka	-	5
Unknown poka	-	5
Mazrapoka	Stem Borer	4
Ghunpoka	Fruit Fly	3
Kath poka	Psyllid Gall	3
Sabujpoka	-	3
Urho poka	-	3
Bichepoka	Hairy Caterpillar	2
Chhaipoka	Jassids	2
Chosokpoka	Thrips	2
Kalopinpre	Black Ant	2
Ui poka	-	2
Pinperhe	Ant	2
Bel poka	-	1
Carentpoka	Trunck Borer	1
Chitanopoka	Mango Fruit Fly	1
Ghugri	Mealy Bug	1
Gutipoka	Red Pumpkin Bettle	1
Karla poka	-	1
Katuipoka	Rhino CerousBettle	1
Kora	-	1
Krimijatiya	Nematode	1
Lal maker	Red Mite	1
Nara poka	-	1
Sumnipoka	Mango Hopper	1

Source: Primary Survey

to different kinds of diseases. The common diseases in our sample home gardens are *Kutha*(41% of households), *Dhasa* (37% of households), *Gara Pacha*(32% households). Some home gardeners report that few species of plants like mango and jack fruit are becoming extinct in the home gardens. The plants which are becoming extinct with the corresponding percentage frequencics are found from primary survey. The major reasons that have been put forward by the households for extinction are felling of trees for making houses, crisis of water, attack of pests, deteriorating quality of the soil in home gardens. Certain animal species are reported extinct. The main reasons for extinction of animals as reported by the households are shortage of food, disease, problem of rearing, inapt environment, shortage of pond etc. In Table 4, we have presented percentage of households reporting different types of insects and pests attacking plants and crops in home gardens. In Table 5, we have presented percentages of households reporting different types of diseases by different plants. In Table 6, percentage of households for stages of the crop/tree infested by pest has been given. In Table 7, damages of the crops for pest infestation have been reported. In Table 8, Control techniques used by the households have been reported. In Table 9, different diseases affecting crops are given. In Table 10, crops and plants affected by diseases have been reported. In Table 11, a list of unusual changes of crops and plants like falling of leaves, immature fruiting that haven been observed in home gardens have been given. In Table 12, changes in flowering and fruiting are given. In Figures 6 to 10 we have given the picture of different diseases of plants and animals found in home gardens.

PESTS AND DISESESES OF POULTRY AND ANIMALS IN HOME GARDENS

Almost all the households in the Home Gardens surveyed rear domestic animals. The diseases found in animals and poultry may be classified as follows: (a) Bacterial, Mycoplasmal, Viral, Fungal, Parasitic, Nutritional, Skeletal Disorder, Kidney Disease and Miscellaneous Disease, (b) Colibesillosis, Yolk-sack Infection, Avian Paste Unrellosis and Others, (c) Ranikhet Disease, Infections Bursal Disease. It is found from the survey that 73% of households have reduced the number of the animals in their home garden, whereas 13% of the home gardeners reported that they increased the number of animals. In Table 13, different pests affecting the animals and poultry in home gardens have been reported. The most prevalent pests are found as ticks and lice. In Table 14, different diseases affecting the animals and poultry in home gardens have been given. The most commons diseases are Ticks, H.S, P.P.R etc. In Table 15, the animals affected by diseases have been given. Cows are found most

Table 5. Crops and Trees Attacked by Insects and Pests

Name of the Crops and Trees	English Name	No. of HHs Reported
Lanka	Chilli	43
Aam	Mango	39
Begun	Brinjal	39
Lau	Gourd	22
Kantal	Jackfruit	18
Kumrho	Pumpkin	8
Narikel	Coconut	8
Sajna	Morunga	8
Pui	Kind of leafy vegetable	7
Lebu	Lemon	5
Payara	Guava	5
Sim	Country bean	5
Batabilebu	Shaddock	4
Karla	Balsam Apple	4
Tomato	Tomato	4
Alu	Potato	3
Jhingre	Cucerbitaceous	3
Chichinga	Snake gourd	2
Jam	Jamboline	2
Kachu	Yam	2
Kul	Indian jujube	2
Kundri	Terner	2
Leafy veg.	Leafy vegetable	2
Anarash	Pine apple	1
Bel	Golden apple	1
Chalta	Elephant apple	1
Chalkumro	A kind of Gourd	1
Dhan	Paddy	1
Golap	rose	1
Kala	Banana	1
Lal sak		1
Mehagini	Mohagany	1
Simul	Kapoke	1
Siris	Shirsa	1
Suo chalta	One type of timber tree	1
Tagar	Kind of flower	1

Source: Primary Survey

Table 6. Stage of the Crop/Tree Infested by Insect/ Pest

Items	No. of HHs
Reproductive	71
Vegetative	65
Seedling	40
Other	2

Source: Primary Survey

Table 7. Damage of the Crop and Tree after Infested by Insect/Pest in Home Gardens

Items	No. of HHs
Yield reduction	90
Quality of yield reduction	49
Complete crop failure	12
Other	3

Source: Primary Survey

susceptible to diseases. In figure 8 to figure 12, we have shown some diseases found in home gardens under survey.

MANAGEMENT OF DISEASES OF PLANTS AND ANIMALS IN HOME GARDENS

Home garden owners face different types of problems in maintaining home gardens including different diseases in the plants and animals there. In Table 16, different problems faced by the owners of the home gardens have been given. The major

Table 8. Control Techniques to Remove Insect/Pest in Home Gardens

Items	No. of HHs Reported
Spray pesticide	26
Dithen M-45 spraying	20
Using pesticide as endocil	9
Using pesticide as feradon	9
Cutting the plant	2

Source: Primary Survey

Table 9. Disease Attracting Crops and Trees

Name of the Disease	English Name	No. of HHs Reported
Kutha rog	Mosaic	42
Dhasa	Downy Mildew	34
Gora pacha	Foot Rot	30
Karkat	Canker	4
Dagapacha	-	4
Blight	Blight	3
Dagasukno	-	3
Patahalud rog	-	3
Patajhara rog	-	2
Halud rog	-	1
Jhalsano	Leaf Blight	1
Jhimano	-	1
Nalipoka	-	1
Panama rog	Panama Disease	1

Source: Primary Survey

problems of home gardens are crisis of water, spoilage of crops and plants by elephants, increasing pests and insects. In Table 17 and Table 18, various types of changes made in the home gardens for the plants and animals have been presented. The major changes made for decreasing cultivation in home garden, cultivating efficiently, more uses of fertilizer etc. In Table 18, we find that 16% of the households have stopped rearing and 14% of the households have decreased rearing of animals in home gardens.

In Table 19, suggestions made by the owners of home gardens to overcome the various problems have been reported.

CONCLUSION

The article presents the field survey based observations on possible contribution of the home gardens in a dry zone in West Bengal in India in changing climate and pests and disease threats to plants and livestock. The easiest and most economical way to manage plant disease in the home garden is by pro-active and scientifically planned maintenance of healthy plants. Successful disease management needs to emphasize monitoring of accurate scientific diagnosis of the problem. Knowing

Table 10. Crops and Trees Attracted by Diseases

Name of the Crop/Tree	English Name	No. of HHs Told
Lanka	Chilli	54
Begun	Brinjal	35
Aam	Mango	22
Lau	Gourd	20
Kanthal	Jackfruit	10
Lebu	Lemon	8
Kumrho	Pumkin	6
Pui	Kind of leafy vegetable	6
Peyera	Guava	5
Sajna	Morunga	5
Karla	Balsam Apple	4
Sim	Country bean	4
Batabi	Shaddock	3
Chalkumra	A kind of gourd	3
Kul	-	3
Narikel	Coconut	3
Tometo	Tomato	3
Chichingga	Snake gourd	2
Jam	Coriander seed	2
Kachu	Yam	2
Kundli	Terner	2
Aalu	Potato	1
Anarash	Pine apple	1
Bel	Golden apple	1
Dhan	Paddy	1
Full gach	Kind of leafy vegetable	1
Jhinge	Cucerbitaceous	1
Kala	Banana	1
Lal sak	A kind leafy vegetable	1
Mehagini	Mohagany	1
Motor	Pea	1
Siris	Shirsa	1
Suochalta	One type of timber tree	1

Source: Primary Survey

Table 11. Nature of the Unusual Change Observed in Home Gardens

Change Observed	No. of HHs
Falling down of leaves	2
Leaves rolling & falling down	2
Suddenly leaves fall	2
Suddenly trees becoming dry	2
Leaves becoming dry	1
Burning of leaves	1
Decreasing growth	1
Drying of leaves	1
Leaves become red & falling down	1
Leaves becoming yellow & falling down	1

Source: Primary Survey

Table 12. Changes in Flowering/Fruiting of Plants during last 20 years

Changes Noticed	No. of HHs Reported
Falling of flowers& fruits	37
Falling of flowers	30
Falling of fruits	20
Buds falling	9
Bud shed	4
Immature fruit drops	3
Fruits becoming small	2
Decrease of fruiting	1
Fruits getting dry	1
Hole in fruits	1

Source: Primary Survey

the common diseases of individual plants may help to identify the disease. The identification of other diseases requires microscopic examination of diseased tissue or more sophisticated laboratory techniques. There is need for increased science-policy connections and extension activities and awareness building among home garden owners for advanced management of home gardens and to be able to cope with the climate change.

Table 13. Pests Affecting Animals in Home Gardens

Name of the Pest	English (Name)	No. of HHs
Atul poka	Ticks(Ectoparasyte)	49
Ukun	Lice	21
Chulchulipoka	-	13
Athalipoka	-	12
Kentarpoka	-	4
Krimipoka	Tape/Round worm	4
Machipoka	-	4
Poka	-	4
Lerhapoka	-	2
Near poka	-	2
Anchul	-	1
Chai poka	Flea	1
Cham ukun	Lice	1
Chhedukun	Lice	1
Chhotoukun	-	1
Dust machi	-	1
Idali	-	1

Source: Primary Survey

ACKNOWLEDGMENT

The authors thankfully acknowledge the financial support from the Asia Pacific Network (APN) for Global Change Programme of Jadavpur University under the three-country research project entitled 'Vulnerability of home garden eco-systems to climate change and its impacts on food security in South Asia' (Project no. ARCP2010-03CMY-Marambe). We also acknowledge Mr. Shyamal Kumar Bandyopadhyay, Mr. Amit Das, Mr. Amitava Chakraborty, Mr. Siddhartha Manna for helping in conducting the study.

Table 14. Diseases Affecting Animals in Home Gardens

Name of the Disease	English Name	No. of HHs Reported
Ausa	Ticks(Ectoparaclyte)	69
Sannipat	Typhus	36
Diarrhoea	P.P.R	24
Jar	Fever/EPF	18
Jhimuni	Drousiness	14
Tunki	R.D	14
Mukhegha	Tounge cancer	9
Pox	Cow pox	7
Pet fola	Tyus pany	6
Amasay	-	3
Gola fola	Battle Jaw	3
Anthrax	-	2
Gachgachi	B.Q	2
Alsa	-	1
Babaja	B.Q(Black Quarter)	1
Basanta	Pox	1
Cancer	Uterine cancer	1
Kane gha	Ear infection	1
Kankalo	Yoke gall	1
Lala nigarman	-	1
Mirgi	epilepsy	1
Patalapykhana		1

Source: Primary Survey

Table 15. Animals Affected by Disease in Home Gardens

Name of the Animal	No. of HHs
Cow	83
Fowl	38
Goat	31
Others	3

Source: Primary Survey

Table 16. Main Problems/Constraints in Maintaining the Home Gardens

Problems	No. of HHs
Crisis of water	57
Crisis of finance/fund for maintenance of HGs	38
Spoil of crops, plants by elephants	20
Increasing incidence of pests & insects	10
Garden hampered by cows, goats	9
Deteriorating quality of soil	6
Shortage of fertilizer & pesticide	4
Crops plants are attacked by diseases	3
Goats and cows eating small trees/pants in HGs	3
Problems of rearing of animals	3
Scarcity of new variety seeds	3
Shading by the trees	3
Water/Irrigation problem in summer	2
Leaves drop on other better house garden so quarreled	2
Maintenance problem of plants	2
Fencing problem	2
Affected by insects	1
Alkalinity of the soil	1
Died plant for more temperature	1
High price of fertilizers	1
Low growth of plants	1
More droughts	1
Security problems in home gardens	1
Shortage of time for maintenance of home gardens	1
Soil erosion	1
Theft problem	1

Source: Primary Survey

Table 17. Changes made in regard to Crops & Plants in home gardens during the last 20 years

Changes Made	% of HHs
Decreasing cultivation in home garden	33
cultivating efficiently	24
More uses of fertilizer	15
Increasing cultivation	12
Cultivated high yielding variety seeds	9
Using pesticide	7
Change in time of planting	5
Change of variety plants	5
Changing type of fertilizer	3
Efficiently using fertilizer	3
Not cultivating in home gardens	2

Source: Primary Survey

Table 18. Various Types of Changes made for the Animals by Households in home gardens during the last 20 years

Changes Made	% of HHs Made Changes
Now not rearing	16
Decreasing rearing	11
Increasing rearing	5
Now rearing	3
Vaccinating	2
Buying high breed bullock	1
Changing feeding process	1
Rearing breed	1
Vaccinating and uses antibiotic	1

Source: Primary Survey

Table 19. Suggestions to Overcome Various Types of Problems/Constraints in Home Gardens

Suggestions	No. of HHs
Arrangement of big water tanks, pump, deep tube well, *bundh* by the government	39
Financial Support for home gardening	20
Financial help by the government	14
Better system of fencing around the HG	13
More use of pesticide	7
Medicine spraying	5
Arrangement of irrigation system	4
More active role of Gram Panchayat (Village Council)	3
Providing pesticide by the government	2
Providing hybrid tree by the government	2
Supplying variety of seeds by the government	2
Arrangement of *sajaldhara prakalpa (Piped drinking water system)*	1
Arrangement of soil testing by the government	1
Improvement of irrigation system	1
Checking soil erosion	1
Subsidy from government	1

Source: Primary Survey

REFERENCES

Weerahewa, J., Pushpakumara, G., Silva, P., Daulagala, C., Punyawardena, R., Premalal, S., ... & Marambe, B. (2012). Are homegarden ecosystems resilient to climate change? An analysis of the adaptation strategies of homegardeners in Sri Lanka. *APN Science Bulletin*, *2*(2), 22–27.

Attygalle, R. (1996). Contribution of home garden tree species towards family nutrition – case study, In H. P. M. Gunasena (Ed.), *Proceedings of the Forth National Workshop on Multipurpose Trees: Research and Development*, pp. 129–143, Kandy, Sri Lanka, April 1-3, 1994.

Ban, N., & Coomes, O. T. (2004). Home Gardens in Amazonian Peru: Diversity and Exchange of Planting Material. *Geographical Review*, *94*(3), 348–367. doi:10.1111/j.1931-0846.2004.tb00177.x

Blaylock, J. R., & Gallo, A. E. (1983). Modeling the Decision to Produce Vegetables at Home. *Journal of Agricultural Economics*, *65*(4), 722–729. doi:10.2307/1240460

Ciftcioglu, G. C. (2017). Social preference-based valuation of the links between home gardens, ecosystem services, and human wellbeing in Lefke Region of North Cyprus. *Ecosystem Services*, *25*, 227–236. doi:10.1016/j.ecoser.2017.05.002

Cromwell, E., Cooper, D., & Mulvany, P. (1991) Agricultural Biodiversity and Livelihoods: Issues and Entry Points for Development Agencies.

Daulagala, C., Weerahewa, J., Marambe, B., Pushpakumara, G., Silva, P., Punyawardena, R., ... Jana, S. (2013). Socio-economic characteristics of farmers influencing adaptation to climate change: Empirical results from selected homegardens in South Asia with emphasis on commercial orientation. *Sri Lanka Journal of Advanced Social Studies*, *2*(2). doi:10.4038ljass.v2i2.6208

Davis, J. R., Brownson, R. C., & Garcia, R. (1992). Family pesticide use in the home, garden, orchard, and yard. *Archives of Environmental Contamination and Toxicology*, *22*(3), 260–266. doi:10.1007/BF00212083 PMID:1616309

De Zoysa, M., & Inoue, M. (2014). Climate change impacts, agroforestry adaptation and policy environment in Sri Lanka. *Open Journal of Forestry*, *4*(5), 439–456. doi:10.4236/ojf.2014.45049

Hitinayake, H. M. G. S. B., De Costa, W. A. J. M., & Jayaweera, K. G. D. (1996). Food trees in multi-layered home gardens in different Agroecological Zones of Kandy District, In H. P. M. a Gunasen (Ed.), *Proceedings of the Seventh National Workshop on Multipurpose Trees: Multipurpose Trees for Food Security*, pp. 252–264. Kandy, Sri Lanka, November 5, 1999.

Jana, S., Roy, J., Marambe, B., Weerahewa, J., Pushpakumara, G., Silva, P., ... Premalal, S. (2015). Home gardens in the Paschim Medinipur District of West Bengal in India: A land use system with multiple benefits. *International Journal of Environment and Sustainable Development*, *14*(2), 191–206. doi:10.1504/IJESD.2015.068606

Jana, S. K., & Roy, J. (2020). Biodiversity and Impacts of Climate Change in Home Gardens: Evidence From a Study in West Bengal, India. In Current State and Future Impacts of Climate Change on Biodiversity (pp. 113-134). IGI Global.

Lebel, L., Rattanawilailak, S., & Lebel, P. (2010). Lazy Gardens – A Sustainability Experiment in the Uplands of Northern Thailand, Draft Working Paper 2, Unit for Social and Environmental Research. Faculty of Sciences, Chiang Mai University, Thailand.

Little, E., Langston, D., & Eaker, T. (2019). *Disease Management In The Home Vegetable Garden,* UGA Cooperative Extension Circular 862, Accessed on July 15, 2019.

Marambe, B., Weerahewa, H. L. J., Pushpakumara, D. K. N. G., Silva, G. L. L. P., Punyawardena, B. V. R., Premalal, S., Miah, G., & Roy, J. (2012). Vulnerability of Home garden Systems to Climate Change and its Impacts on Food Security in South Asia, Project Report. *Submitted to APN, (Project No. ARCP2010-03CMY-Marambe)*. Retrieved from www.apn-gcr.org/.../files/3ddf57b875774091a38f95bd fee0b6c9.pdf, accessed on September 10, 2012.

Mattsson, E., Ostwald, M., Nissanka, S. P., & Marambe, B. (2013). Homegardens as a multi-functional land-use strategy in Sri Lanka with focus on carbon sequestration. *Ambio*, *42*(7), 892–902. doi:10.100713280-013-0390-x PMID:23456780

Nguyen, T. (1997). A Study on the Home Garden Ecosystem in the Mekong River Delta and the Hochiminh City, Working Paper No. 7, UNESCO, Division of Ecological Sciences, South South Co-operation Programme. Retrieved from http://www.unesco.org/mab/activity/s-s/s-home.html

Pushpakumara, G., Marambe, B., Silva, G. L. L. P., Weerahewa, J., & Punyawardena, R. (2012). Homegardens in Sri Lanka: The status importance and future perspective. *Tropical Agriculturist*, *159*, 55–125.

Saikia, P., Choudhury, B. I., & Khan, M. L. (2012). Floristic composition and plant utilization pattern in home gardens of Upper Assam, India. *Tropical Ecology*, *53*(1), 105–118.

Smith, T., & Weeks, E. (2019). Disease Management In The Home Vegetable Garden Agriculture And Landscape Program 4/09, Accessed on July 15,2019.

Tangjang, S., & Arunachalam, A. (2009). Role of traditional home garden systems in North East India, *Indian Journal of Traditional Knowledge, January, 8*(1), pp. 47–50.

KEY TERMS AND DEFINITIONS

Biodiversity: It is short form of biological diversity. It indicates the totality of genes, species and ecosystems in a region or the world.

Climate Change: According to IPCC, Climate change refers to a change in the state of the climate that can be identified (e.g., by using statistical tests) by changes in the mean and/or the variability of its properties and that persists for an extended period, typically decades or longer.

Ecosystem: A dynamic complex interaction of biotic and abiotic components of environment in a given region.

Gram Panchayat: Gram panchayat is a local body at village level formed under Panchayat Raj Act.

Home Garden: "Home Garden" is a complex multifunctional land system that combines multiple farming components of the homestead and provides environmental services, household needs.

Pest: A destructive insect which attacks crops, food, livestock, etc.

Woody Trees: A plant with thick and tough stems.

APPENDIX: DIFFERENT DISEASES IN THE PLANTS IN HOME GARDENS

Figure 8. Disease banana leaves

Figure 9. Kutha disease

Figure 10. Disease in Rose

Figure 11. Disease in Jamrul

Figure 12. Disease in Brinjal

Section 2

Climate Change and Neglected Tropical Diseases Associations

Chapter 4
Network Modelling on Tropical Diseases vs. Climate Change

G. Udhaya Sankar
https://orcid.org/0000-0002-1416-9590
Alagappa University, India

C. Ganesa Moorthy
https://orcid.org/0000-0003-3119-7531
Alagappa University, India

ABSTRACT

This chapter has proposed a systematic method to design mathematical models. These models have been associated with counting of white blood cells, counting of red blood cells, population of mosquitoes, and counting of foreign bodies like virus, bacteria, and parasite in a human body. Interpretations for critical points or equilibrium points have been given for network systems of differential equations associated with models. A practical method of applying these interpretations in administrating medicines to get control over diseases has been given. Order of priority in three types of critical points, namely, stable critical points, unstable critical points, and asymptotically stable critical points, has been given. Conversions of differential equations of models into integral equations and applying Picard's iteration method to solve integral equations have been explained. A step-by-step approach has been used in designing models, solving models, and interpreting solutions of models for tropical diseases.

DOI: 10.4018/978-1-7998-2197-7.ch004

INTRODUCTION

Tropical countries are the countries which are near to the equator of our earth planet, which are supposed to have high temperature, which are homelands for many insects which spread diseases, and which have many diseases solely like malaria, dengue, chikungunya etc. There is a chance to have hot conditions and humid conditions in these regions and these conditions are good enough for infectious diseases. Clinical tests and experimental tests help to understand nature of diseases and nature of medicines. But, they are not sufficient to explain many things, because results of tests in one region may contradict results of tests in another region. Only mathematical models, equations and solutions can explain some more things, more specifically about confronting medical facts. But these models alone cannot explain everything, because models cannot be complete in all aspects, and parameters involved in models can be found only by means of experimental data. Even in clinical tests, one has to depend on statistical methods in terms of measures of central tendency, measures of dispersions, correlation coefficients, regression lines, estimations and hypotheses testing which depend on probability. Regression line methods are modified as curve fitting, by guessing the curves in terms of solutions of equations involved in models. Probabilistic methods are modified as stochastic methods to understand long term effects. There are articles in literature for all these things explained above, and researches continue, because tropic countries are most affected countries by climate changes which happen.

The purpose of this chapter is to develop procedures to design models (Figure. 1) for tropical diseases and explain tools to analyze these models. It all begins with known prey-predator equations just to understand the beginning of the art of designing a model. Without correct justifications the words prey and predator are changed to the words foreign bodies and white blood corpuscles just to explain suitable mathematical methods to solve differential equations involved in the model and to guess suitable curves for curve fitting from forms of solutions of differential equations. Picard's iteration method is considered as the most promising method for solving differential equations in this chapter. Synchronization methods are considered as the most favorable methods when the models are dynamic ones. Then an exact simplified model for foreign body-white blood corpuscle is designed. Complications are considered in terms of variations in temperature due climate changes, in designing models. A model is designed for network relations connecting growth of mosquito-population, white blood cell-population (counting), and malaria parasite-population. A model is designed for relations of numbers of malaria parasites, white blood cells, and red blood cells which would be helpful in understanding interrelationship between anemia and malaria. All these things are

Figure 1. Tropical Diseases vs. Climate Change

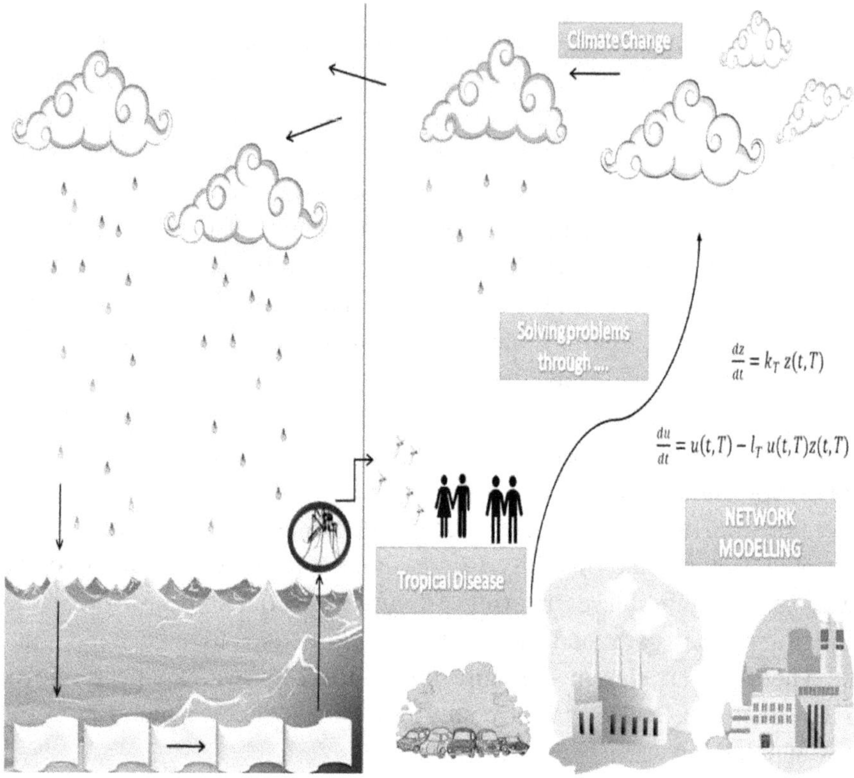

towards exact interpretations for equilibrium points and their stability. Some possible interesting theoretical conclusions are derived in this chapter.

MODELS FOR ENCOUNTERS OF FOREIGN BODIES WITH WHITE BLOOD CORPUSCLES

Background

There are many tropical diseases like: Chagas disease, Dengue, Helminths, African trypanosomiasis, Leishmaniasis, Leprosy, Lymphatic filariasis, Malaria, Onchocerciasis, Schistosomiasis, Hookworm, Trichuriasis, Treponematoses, Buruli ulcer, Dracunculiasis, Leptospirosis, Strongyloidiasis, Foodborne trematodiases, Neurocysticercosis, Scabies, Flavivirus infections, etc. There are many articles, Greenwood, M. (1916), Kermack, W. O. et al, (1927), Kermack, W. O. et al., (1932), Kermack, W. O. et al., (1933), Anderson, R. M., (1988), Hethcote, H. W. (2000),

Bernoulli, D.et al., (2004), Eubank, S. et al., (2004), Keeling, M. J. et al., (2005), Shirley, M. D. et al, (2005), Chitnis, N. et al., (2018), Gervas, H. E. et al., (2018), Rock, K. S. et al., (2018), Bañuelos, S. et al., (2019), Chowell, G.et al., (2019), Musa, S. S. et al., (2019), which provide mathematical approaches to analyze diseases to enable us to get a control over diseases.

MAIN FOCUS OF THE CHAPTER

Our ultimate aim is to control diseases. This is done experimentally as well as mathematically, and both of them are necessary ones. Experimental results may give present status, but not more details. But mathematical models can help us to guess some information for future status. However, some undetermined factors in mathematical models can be determined only by means of experimental observations. So, both of them are necessary ones. This chapter focuses only on mathematical methods. If there are some nodes or junctions in an algorithm, then the algorithm is a network algorithm. Sometimes mathematical models may also have nodes or junctions, and in this case these models are called network models. Mostly all algorithms are network algorithms and all models are network models. Models for encounters of virus (or bacteria or parasite) with white blood corpuscles are network models, and these models have been described in this section of the chapter. Some unusual (but) interesting mathematical methods like power series method and Picard's iteration method are applied for mathematical procedures. Synchronization methods are refinements of known mathematical methods. Refinements of Picard's methods for synchronization methods in discussing tropical diseases like heat rashes have been discussed in this chapter. The medical interpretation for equilibrium points or critical points for systems of ordinary differential equations meant for encounters between foreign bodies and white blood corpuscles are provided in this section of the chapter, and it has been justified that critical points are the most wanted points required for disease control. Among the critical points there is a classification: stable points, asymptotically stable points, unstable points. It is pointed out through medical interpretation that asymptotically stable critical points are the most wanted points. Basic information which can be derived from these concepts for practical implementation is provided. For example, the following known practice is derived. Suitable time breaks should be given between two successive intakes of medicines.

Figure 2. Populations of foxes and rabbits

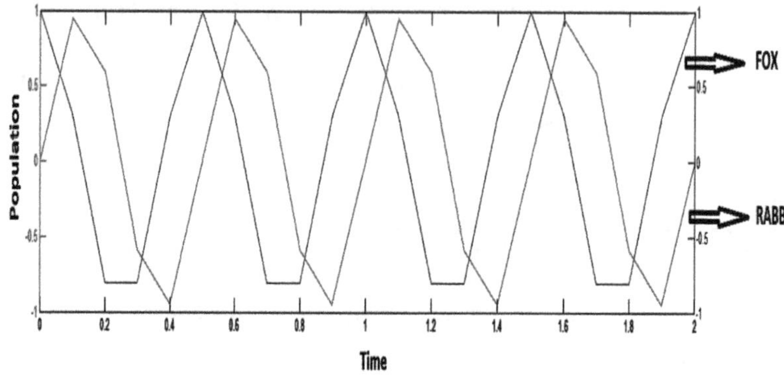

PREY-PREDATOR EQUATIONS

Volterra (1860-1940) was an Italian mathematician who worked in application oriented pure mathematical analysis and in mathematical biology. One of his works is famous Volterra's prey-predator equations. One way of explaining these equations is the following way. Let us imagine that there are foxes (predators) and rabbits (preys) in an island. There are abundant clovers in the island which are continuous food for rabbits. But, rabbits are food for foxes. If the foxes are too many in number in a period, they eat too many rabbits and the population of rabbits begins to decline. When the population of rabbits is reduced in number, then many foxes do not get enough food and population of foxes reduces because of starvation. Because of reduction in the population of foxes, the rabbits are relatively safe and the population of rabbits increases. Since the population of rabbits increases, again foxes get abundant food and the population of foxes increases. So, there are endless cycles like the ones described in the Figure 2.

Let us now do *modelling* for this *network* connecting population of foxes and population of rabbits, mathematically. Let $x(t)$ or simply x denote the number of rabbits at a general time t. Let $y(t)$ or simply y denote the number of foxes at time t. Then there are constants $a > 0$ and $b > 0$ such that

$$\frac{dx}{dt} = ax - bxy,$$

where ax is a part of increase in the rate of change of population $\frac{dx}{dt}$ depending on direct proportion of the present population x at time t, and where $-bxy$ is a part of decrease in the rate of change of population $\frac{dx}{dt}$ depending on direct proportion of the present possible encounters xy between x number of rabbits and y number of foxes at time t. Similarly, there are positive constants c and d such that

$$\frac{dy}{dt} = -c\,y + d\,xy\,,$$

for which an interpretation can be given. Thus there are following nonlinear equations to describe the changes in populations:

$$\frac{dx}{dt} = x\left(a - b\,y\right), \tag{1}$$

$$\frac{dy}{dt} = -y\left(c - d\,x\right). \tag{2}$$

These equations (1) and (2) are called Volterra's prey-predator equations. This model is a simplified model without complicated restrictions. One can consider additional simple restrictions. It may happen that the food clover for rabbits may be restricted, for example a *climate change* in the island may affect the growth of clovers and the cycles described above may be affected. A climate change may directly affect the population of foxes, and again there may be changes in cycles.

FB-WBC EQUATIONS

Let us now change the words for another interpretation of this model. Let us change the word "island" by "human body", "rabbits" by "foreign bodies" like virus and bacteria, "foxes" by "white blood corpuscles", and "clovers" by "human body cells". Then two equations of the type (1) and (2) are obtained, where $x\left(t\right)$ denotes the number of a particular type of foreign bodies (FBs), and $y\left(t\right)$ denotes the number of white blood corpuscles (WBCs). One may object to use the term $-cy$ in the equation (2) for this conversion of words, because WBCs are produced from stem cells, and in that case it may be assumed that $c = 0$. One may also object $d > 0$ in the equation (2) for this conversion of words, and in that case it may be assumed

Figure 3. Populations of FBs and WBCs

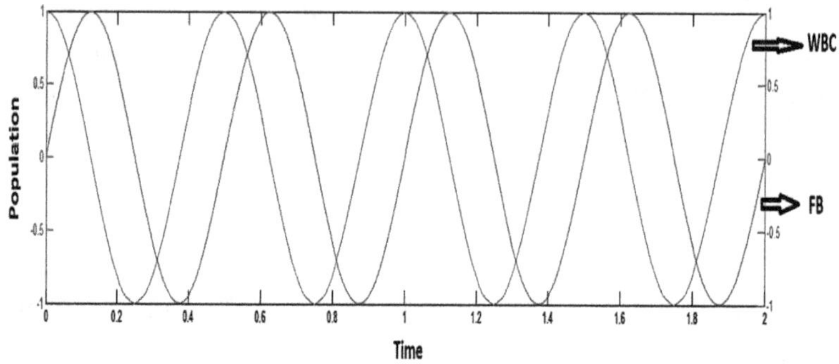

that $d < 0$, because these conditions will not be considered for our discussions at present. However, an exact modification will be done later.

These two (types of) equations (1) and (2) are sufficient for representation in simplified environments. One should consider maximum possible number of WBCs in a counting, change in a counting of FBs corresponding to a change in body temperature, maximum and minimum possible temperatures of a human body, minimum number of WBCs required being alive, etc. Let us call such restrictions as additional restrictions in the remaining part. Thus, in general, there are two major first order nonlinear ordinary differential equations along with some auxiliary differential equations or inequalities. If only two major equations are considered, then there are some simple procedures to solve them, and if auxiliary equations are also considered, then there are some complicated procedures to solve them. By solving them, it is possible to find $x(t)$ and $y(t)$ at a general time t. What is the purpose of finding $x(t)$ and $y(t)$? The ultimate aim is to get a control over diseases or FBs; not just getting solutions. The next discussion will be towards this aim.

Let $x = x(t)$ and $y = y(t)$ be solutions of the equations

$$\frac{dx}{dt} = F(x, y), \tag{3}$$

$$\frac{dy}{dt} = G(x, y), \tag{4}$$

when (1) and (2) are of these forms. Here $F(x, y)$ and $G(x, y)$ are fixed functions of x and y. Eliminate t from $x = x(t)$ and $y = y(t)$ to obtain a curve in the form $H(x,y)=0$. There may be some special points at which t cannot be eliminated. For example, if (x_0, y_0) satisfies $F(x_0, y_0)=0$ and $G(x_0, y_0)=0$, and if there is a t_0 such that $x(t_0)=x_0$ and $y(t_0)=y_0$, then $\frac{dx}{dt}=0$ and $\frac{dy}{dt}=0$ at this point and hence the solutions curves (like the ones given in Figure 3) should be locally constant functions and hence elimination at t_0 apparently fails. Such points (x_0, y_0), where elimination is not possible, are called critical points or equilibrium points. The term "equilibrium points" is used because the intersecting solution curves (see Figure 3) suggest that $x(t_0)=y(t_0)$ at this special point. That is, the number of FBs is equal to the number of WBCs capable to encounter FBs. This situation is a required condition in a human body to get a control over diseases. So, it is always required to find critical points or equilibrium points to get a control over diseases. Although a formal definition for critical points has been given, some more explanations are required.

If $x=x(t)$ and $y=y(t)$. are solutions of (3) and (4) and c is any constant, then $x=x(t+c)$ and $y=y(t+c)$ are also solutions of (3) and (4). Thus, there are infinitely many "$H(x,y)=0$" which can be obtained by varying c. Thus, there are infinitely many solution curves with variables x and y. These curves $H(x,y)=0$ can also obtained from the infinitely many solutions of $\frac{dx}{dy}=\frac{F(x,y)}{G(x,y)}$, which can be derived from (3) and (4). So, the solution curves $H(x,y)=0$ are varied, and a suitable curve with variables x and y can be found such that this curve passes through any given point, say, (x_0, y_0). But, if our chosen point (x_0, y_0) is a critical point, then it is difficult to find a curve $H(x,y)=0$ passing through (x_0, y_0), in view of the previous discussion. However, if this equilibrium point is isolated in the sense that there is no other critical point in a neighborhood of this equilibrium point, then one can find curves $H(x,y)=0$ in terms of x an y in that neighborhood. These curves which are close to an equilibrium point classify the nature of this point in three types. Let (x_0, y_0) be a critical point. Let $(x(t), y(t))$ define a parametric solution curve. The critical point (x_0, y_0) is said to be stable, if for given $R>0$, and if one point of the solution curve $(x(t), y(t))$ enters in the disc D with centre (x_0, y_0) and radius R, then it implies that the entire curve lies in the disc D. The critical point (x_0, y_0) is said to be asymptotically stable, if it is stable and if, for

Figure 4. Three types of critical points

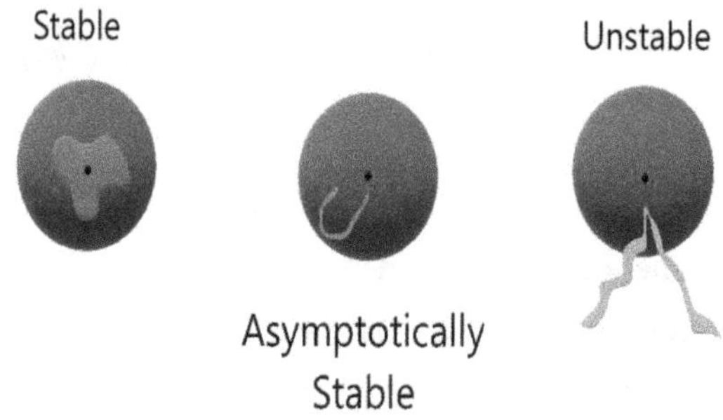

given R and D mentioned above, $\left(x\left(t\right), y\left(t\right)\right)$ approaches $\left(x_0, y_0\right)$ as t approaches a limit of parametric values. The critical point $\left(x_0, y_0\right)$ is said to be unstable, if it is not stable.

In a FB-WBC model, it is required that the human body should necessarily reach an equilibrium point to get a control over a disease, because it is expected that $y\left(t\right) \geqslant x\left(t\right)$ at some time t for a recovery from a disease, and for this purpose the body should reach first $x\left(t_0\right) = y\left(t_0\right)$ for some time t_0. So, it is desirable to find critical points. Among the critical points most desirable are asymptotically stable critical points, then next desirable are stable points, and then least desirable are unstable points.

SOLUTION AND RECOMMENDATIONS

- Depending on the nature of a disease and the nature of medicines, undetermined constants or undetermined functions F and G involved in the equations (3) and (4) (for example) should be found by using experimental observed values to fix the equations (3) and (4), specifically.
- A critical point $\left(x_0, y_0\right)$ should be found with the following order of priority: asymptotically stable critical point; stable critical point; unstable critical point. In some occasions, it would be preferable to select this point such that x_0 is near to the present value of x in a human body.
- Present $x\left(t\right)$ should be found. If it is not near to x_0, it should be controlled by medicines to bring the value of $x\left(t\right)$ neat to x_0.

- A break should be given in providing medicines for a period (in minutes/ hours/ days/ weeks…). This break may change the values of $y(t)$ as well as $x(t)$.
- Since $x(t)$ gets a deviation from x_0, medicines should be used again to bring the value of $x(t)$ near to x_0.
- Thus, medicines should be given with suitable breaks. The break-periods in practice depend on nature of the human body, nature of existing diseases, and nature of medicines used.

SOLUTIONS BY MATHEMATICAL METHODS

There are many known methods to solve (3) and (4). Power series methods are applicable to solve some special equations like (1) and (2). Let us next explain the same.

POWER SERIES METHOD

Let us consider the following equations

$$\frac{dx}{dt} = ax + bxy + c \tag{5}$$

$$\frac{dy}{dt} = dy + exy + f \tag{6}$$

with the initial conditions $x(0) = a_0$ and $y(0) = b_0$, where $a, b, c, d, e, f, a_0, b_0$ are constants. Assume that the solutions take the forms

$$x = \sum_{n=0}^{\infty} a_n t^n, \tag{7}$$

$$y = \sum_{n=0}^{\infty} b_n t^n, \tag{8}$$

where the unknown constants $a_1, a_2, \ldots, b_1, b_2, \ldots$ are to be determined. Substitute (7) and (8) in (5) and (6) to obtain

$$\sum_{n=1}^{\infty}\left[\left(n+1\right)a_{n+1} - aa_n\right]t^n + \left(a_1 - aa_0 - c\right) = ba_0b_0 + b\sum_{n=1}^{\infty}\left[\sum_{k=0}^{n} a_k b_{n-k}\right]t^n, \tag{9}$$

and

$$\sum_{n=1}^{\infty}\left[\left(n+1\right)b_{n+1} - db_n\right]t^n + \left(b_1 - db_0 - f\right) = ea_0b_0 + e\sum_{n=1}^{\infty}\left[\sum_{k=0}^{n} a_k b_{n-k}\right]t^n. \tag{10}$$

On comparing the coefficients of t^n in (10), there is a relation

$$\sum_{k=0}^{n} a_k b_{n-k} = \frac{1}{e}\left[\left(n+1\right)b_{n+1} - db_n\right]. \tag{11}$$

This reduces (9) to

$$\sum_{n=1}^{\infty}\left[\left(n+1\right)a_{n+1} - aa_n\right]t^n + \left(a_1 - aa_0 - c\right) = ba_0b_0 + \frac{b}{e}\sum_{n=1}^{\infty}\left[\left(n+1\right)b_{n+1} - db_n\right]t^n. \tag{12}$$

On comparing the coefficients in (9), (10), and (12), there is a complete set of following relations.

$$a_1 = ebb_0 + c + aa_0,$$

and

$$b_1 = ea_0b_0 + f + db_0.$$

For $n \geq 1$,

$$a_{n+1} = \frac{1}{n+1}\left[aa_n + b\sum_{k=0}^{n} a_k b_{n-k}\right],$$

and

$$b_{n+1} = \frac{e}{b}a_{n+1} - \frac{e}{b(n+1)}a_n + \frac{ed}{b(n+1)}b_n .$$

Using this set of relations, one can find many coefficients successively and one can find the sums $\sum a_n t^n$, $\sum b_n t^n$ up to any finite stage so that they provide good approximate (not numerical) solutions. This illustrates the power series method to solve (5) and (6) under the additional assumptions that there are solutions in the form (7) and (8).

PICARD'S ITERATION METHOD

Consider (3) and (4) along with initial conditions $x(0) = a_0$ and $y(0) = b_0$. Solving this problem is equivalent to solving the following integral equations.

$$x(t) = a_0 + \int_0^t F(x(s), y(s))ds . \tag{13}$$

$$y(t) = b_0 + \int_0^t G(x(s), y(s))ds . \tag{14}$$

Define $x_0(t) = a_0$ and $y_0(t) = b_0$, for all t. Define $x_n(t)$ and $y_n(t)$ successively by the following relations.

$$x_{n+1}(t) = a_0 + \int_0^t F(x_n(s), y_n(s))ds$$

and

$$y_{n+1}(t) = b_0 + \int_0^t G(x_n(s), y_n(s))ds .$$

It is expected in general that $x_n(t)$ and $y_n(t)$ converge to $x(t)$ and $y(t)$, which are solutions of (13) and (14), when n tends to infinity. Thus, $x_n(t)$ and $y_n(t)$ are good approximate solutions of (3) and (4), when n is sufficiently large. This method of finding solutions of (3) and (4) is called Picard's method.

For the Picard's method for the equation $\frac{dx}{dy} = \frac{F(x,y)}{G(x,y)}$, along with the initial condition $x(0) = c_0$, consider the equivalent integral equation

$$x(y) = c_0 + \int_0^y \frac{F(x(s),s)}{G(x(s),s)} ds .$$

The corresponding Picard's iteration formula is

$$x_{n+1}(y) = c_0 + \int_0^y \frac{F(x_n(s),s)}{G(x_n(s),s)} ds ,$$

along with the function $x_0(y) = c_0$, for all y.

When n is sufficiently large, convert $x = x_n(y)$ in the form $H_n(x,y) = 0$. Then $H_n(x,y) = 0$ is an approximation to $H(x,y) = 0$, which is a solution of $\frac{dx}{dy} = \frac{F(x,y)}{G(x,y)}$.

CURVE FITTING

If it is guessed from solutions of linearization of (5) and (6) that the solutions of (5) and (6) approximately combinations of the functions of the type $1, t, e^{kt}, \cos kt, \sin kt$, then it may be assumed that approximate solutions of (5) and (6) are of the forms

$$x = c_1 + c_2 t + c_3 e^{c_4 t} + c_5 \cos c_6 t + c_7 \sin c_8 t \qquad (15)$$

and

$$y = c_9 + c_{10}t + c_{11}e^{c_{12}t} + c_{13}\cos c_{14}t + c_{15}\sin c_{16}t, \tag{16}$$

where $c_1, c_2, \ldots, c_{16}$ are constants to be determined. For example, one can make enough observations, and the values may be used to determine $c_1, c_2, \ldots, c_{16}$ by least square principle. One may see, for example the article of Serfling, R. E. (1963), for this type of guessed solutions. Thus there is one more method of finding solutions without solving differential equations, but by guessing solution forms and by making observations.

Apart from all these methods, there are some other standard methods to solve (3) and (4) practically. But, it does not mean that it is possible to have control over all types of diseases, because other parameters and other constraints should also be considered. When other constraints are also considered, the methods mentioned above should be modified. More specifically, "Synchronization" should be done.

SYNCHRONIZATION METHODS

The literal meaning of "synchronization" is making time adjustments so that adjusted times are nearly equal to a particular reference time. The deviations in time should be adjusted. The deviations from constraints should be adjusted in any "synchronization method" in mathematics. Let us discuss a *new* synchronization method based on Picard's method. One person may be a prime person and other persons may be secondary persons, then secondary persons may synchronize their time so that the time coincides with the time of the prime person. Sometimes a second person may adjust his time with that of a first person, then a third person may adjust his time with that of the second person, and then a fourth person may adjust his time with that of the third person and this procedure of synchronization may be continued. These particular chain adjustment type mathematical synchronization methods are used in this chapter.

TROPICAL DISEASES - HEAT RASHES

Let us consider the simple tropical disease "heat rashes" which is common in India during summer seasons, more specifically during the months of April, May, June, and July. The average temperature may vary from 100°F to 120°F during day times in this period. Let us begin with the model of (1) and (2) for this purpose. The parametric constants may depend on temperature. More specifically, the constants a and c may depend on temperature in (1) and (2) in some scale, when (1) and (2) are considered as a model for FBs and WBCs. For example, let us replace a and c in (1) and (2) by $a\left(T\right)$ and $c\left(T\right)$, when T is considered from 1 to 20 with an understanding that heat rashes may appear only at 101°F and it may get increased up to 120°F in India (only for an understanding purpose). For example, if temperature is 111°F, then T gets the value 11 from the difference 111°F-100°F. In general, $a\left(T\right)$ and $c\left(T\right)$ increase with T, and hence $a\left(T\right)$ and $c\left(T\right)$ may be replaced by $Ta\left(1\right)$ and $Tc\left(1\right)$, where $1 \le T \le 20$. Let us again replace $a\left(1\right)$ and $c\left(1\right)$ simply by a and c. Thus one has the following equations.

$$\frac{dx}{dt} = Tax - bxy \, , \tag{17}$$

and

$$\frac{dy}{dt} = -Tcy + dxy \, , \tag{18}$$

for $1 \le T \le 20$ corresponding to temperature variation from 101°F to 120°F, along with initial conditions $x\left(0\right) = a_{0T}$ and $y\left(0\right) = b_{0T}$. Then a simplified set of Picard's iteration formulas are

$$x_{n+1}\left(t\right) = a_{0T} + \int_0^t \left[Tax_n\left(s\right) - bx_n\left(s\right)y_n\left(s\right)\right]ds \, , \tag{19}$$

and

$$y_{n+1}\left(t\right) = b_{0T} + \int_0^t \left[-Tcy_n\left(s\right) + dx_n\left(s\right)y_n\left(s\right)\right]ds \, , \tag{20}$$

for $1 \le T \le 20$. Suppose $T_1, T_2, T_3, \ldots$ are simplified discrete successive different values of T, which is assumed for convenience. Here, for example, the variable t corresponding to $T = T_2$ is the present time by referring to time of change of T from T_1 to T_2 as 0. For $T = T_1$, perform iterations using (19) and (20) with

$$x_0(t) = a_{0T_1}, y_0(t) = b_{0T_1},$$

for all t, for sufficiently large number of times till $x_{n_1}(t), y_{n_1}(t)$ are obtained. Then for $T = T_2$, perform iterations using (19) and (20) with

$$x_0(t) = x_{n_1}(t), y_0(t) = y_{n_1}(t),$$

for all t, for sufficiently large number of times till $x_{n_2}(t), y_{n_2}(t)$ are obtained. Then for $T = T_3$, perform iterations using (19) and (20) with

$$x_0(t) = x_{n_2}(t), y_0(t) = y_{n_2}(t),$$

for all t, for sufficiently large number of times till $x_{n_3}(t), y_{n_3}(t)$ are obtained. Let us proceed in this way. Thus, updating for initial approximations may be done by using the solutions obtained for previous T level. This is one simple modification in using Picard's iteration method while solutions are found at different phases of T. This is a synchronization method. Let us discuss another synchronization method for another required situation, which does exist.

It happens in real situation that there are upper bounds (in an average unit volume) for FBs as well as for WBCs. That is, there are constants K_1 and K_2 such that $x(t) \leq K_1$ and $y(t) \leq K_2$, for all t. Thus, one can now consider the following problem (without considering temperature, for simplicity).

$$\frac{dx}{dt} = ax - bxy,$$

$$\frac{dy}{dt} = -cy + dxy,$$

$$x(t) \leq K_1, \text{ for all } t,$$

$$y(t) \leq K_2, \text{ for all } t.$$

Now, one can follow the following synchronization algorithm with $x_0(t) = a_0$ and $y_0(t) = b_0$, for all t.

$$x_{n+1}^{\sim}(t) = x_0 + \int_0^t \left[a\,x_n(s) - b\,x_n(s)\,y_n(s)\right] ds,$$

$$y_{n+1}^{\sim}(t) = x_0 + \int_0^t \left[-c\,x_n(s) + d\,x_n(s)\,y_n(s)\right] ds,$$

$$x_{n+1}(t) = \min\{x_{n+1}^{\sim}(t), K_1\},$$

$$y_{n+1}(t) = \min\{y_{n+1}^{\sim}(t), K_2\}, \text{ for all } t.$$

Then, $x = x_n(t)$ and $y = y_n(t)$ are approximate solutions, when n is sufficiently large. The corresponding changes can be done when temperature is also considered.

ORIGINS IN MATHEMATICS

There is an expectation for applications for everything in mathematics, just to ensure that only meaningful ideas do exist, which are useful to our human society. On the other hand, when there is a conflict in applied ones, one has to refer to logically correct pure abstract mathematical statements. This does not mean that there is a justification in pure abstract mathematics for every real life application. But all should always develop pure abstract mathematics with an expectation that almost all practical applications would be justified, and there is a need to have abstract part of mathematics related to existing applications. Let us next see that the followings are origins in pure mathematics for justifications of the synchronization methods discussed above. An element x in a nonempty set X is called a *fixed point* of a mapping $f : X \to X$, if $f(x) = x$ is satisfied. This concept can be extended for some more types of non self valued mappings. Let us refer to the book of Kreyszig, E. (1978) for definitions of the concepts to be used in this part, and these definitions are not to be given here. One may also refer to Kreyszig, E. (1978) for the statement of the classical Banach fixed point theorem.

THEOREM: Let (X,d) be a complete metric space. Let Y be a closed subset of X so that (Y,d) is also a complete metric space. Suppose that there is a mapping $g: X \rightarrow Y$ such that

$$d(g(x),g(y)) \leq d(x,y),$$

for all x and y in X, and such that $g(x) = x$, for all x in Y. Let $f: Y \rightarrow X$ be a mapping such that

$$d(f(x),f(y)) \leq kd(x,y),$$

for all x and y in Y, for some constant k satisfying $0 < k < 1$. Then f has a unique fixed point $y*$ in Y. Moreover, for any fixed y_0 in Y, the sequence $(y_n)_{n=1}^{\infty}$ in Y defined successively by $y_n = (g(f(y_{n-1}))$ converges to that unique fixed point.

PROOF: Let $h: Y \rightarrow Y$ be the composition of g and f. Then,

$$d(h(x),h(y)) \leq kd(x,y),$$

for all x and y in Y. By the classical Banach fixed point theorem, $h: Y \rightarrow Y$ has a unique fixed point $y*$ in Y. Fix y_0 and the sequence $(y_n)_{n=1}^{\infty}$ in Y, as mentioned in the statement. By the same Banach fixed point theorem, this sequence converges the unique fixed point $y*$. Moreover, any point in Y is a fixed point of $h: Y \rightarrow Y$, if and only if it is a fixed point of $f: Y \rightarrow X$. This proves that $f: Y \rightarrow X$ also has a unique fixed point. This completes the proof.

This theorem is an abstract origin of the second synchronization method mentioned above. The methodology of converting this theorem towards applications can be found in the book of Kreyszig, E. (1978). The next theorem is an abstract origin of the first synchronization method mentioned above, and one more abstract origin can be found in the article of Moorthy, C. G. et al., (2019).

THEOREM: Let (X,d) be a complete metric space. Let $(f_\alpha)_{\alpha \in I}$ be a collection of mappings from X into itself, such that

$$d\left(f_{\alpha}\left(x\right),f_{\beta}\left(y\right)\right)\leq kd\left(x,y\right),$$

for all $x,y\in X$, for all $\alpha,\beta\in I$, and for some constant k satisfying $0<k<1$. Then, $\left(f_{\alpha}\right)_{\alpha\in I}$ has a common fixed point x^*. Moreover, for any sequence $\left(\alpha_{n}\right)_{n=1}^{\infty}$ in I, for any fixed x_0 in Y, the sequence $\left(x_{n}\right)_{n=1}^{\infty}$ in X defined successively by $x_n=f_{\alpha_n}\left(x_{n-1}\right)$ converges to that unique fixed point.

PROOF: For each n, it is true that

$$d\left(x_{n+1},x_{n}\right)=d\left(f_{\alpha_{n+1}}\left(x_{n}\right),f_{\alpha_{n}}\left(x_{n-1}\right)\right)\leq kd\left(x_{n},x_{n-1}\right)\leq k^{n}d\left(x_{1},x_{0}\right).$$

Thus, for $m>n$, it is true that

$$d\left(x_{m},x_{n}\right)\leq d\left(x_{m},x_{m-1}\right)+d\left(x_{m-1},x_{m-2}\right)+\ldots.+d\left(x_{n+1},x_{n}\right)\leq\left(k^{m-1}+k^{m-2}+\ldots+k^{n}\right)d\left(x_{1},x_{0}\right).$$

This means that the sequence $\left(x_{n}\right)_{n=1}^{\infty}$ in the complete metric space X is a Cauchy sequence, and hence it converges to some x^*. Fix $\beta\in I$. Since f_β is continuous, $\left(f_{\beta}(x_{n})\right)_{n=1}^{\infty}$ converges to $f_{\beta}\left(x^*\right)$. So,

$$d\left(x^*,f_{\beta}\left(x^*\right)\right)=\lim_{n\to\infty}d(f_{\alpha_{n+1}}\left(x_{n}\right),f_{\beta}\left(x_{n}\right))\leq k\lim_{n\to\infty}d\left(x_{n},x_{n}\right)=0.$$

This proves that $f_{\beta}\left(x^*\right)=x^*$. This completes the proof.

Synchronization methods are good promising methods to solve equations of FB-WBC models, more specifically in connection with the tropical disease, heat rashes. On many occasions everyone needs experimental observations in addition to mathematical network modelling methods. However, without using observations one can have some theoretical conclusions. For example, it has been concluded that medicines should be administered with breaks. One can also guess the forms of solutions from theory, even though the undetermined constants should be determined only by means of experimental observations.

MOSQUITOES, TROPICAL DISEASES, AND CLIMATE CHANGES

Among the vector borne diseases, the diseases spread by mosquitoes are brought under tropical diseases because climate changes affect the population of mosquitoes as well as the diseases spread by mosquitoes. There are articles, Githeko, A. K. et al., (2000), Newfield, T. P. (2016), which support that the temperature rise in our world due to climate changes is favorable to growth of population of mosquitoes and thereby there is an increase in tropical diseases. On the other hand, there are articles which support that the temperature rises in our world make a decline in growth of population of mosquitoes and thereby there is a decrease in tropical diseases. Which one is correct? This is classified as an outstanding question in the review article Tjaden, N. B. et al., (2018), which states precisely the following. How can the comparability between different modelling approaches be increased? The reason for raising this outstanding question lies in the difficulty to judge whether all factors are included into consideration when a model is designed. More specifically, when this question is considered for mosquito population, it has to be noticed that there is a possibility to have an increase in the population in one region as well as there is a simultaneous possibility to have a decrease in the population in another region. Because of the climate changes, the temperature raises in our world. So, a cold land becomes a wet land, and there is a chance to have an increase in mosquito population in that land, or region. Simultaneously, a wet land becomes a dry land, and there is a chance to have a decrease in mosquito population in that land. Thus there are two changes in population and thereby decrement as well increment in spread of diseases happen simultaneously, but in different regions because of climate changes. Let us discuss a simple model in connection with climate changes and spread of diseases in terms of population of mosquitoes. This is to be done based on our earlier discussion of this chapter.

MODELLING

This modelling procedure is simplified by the assumption that the mosquitoes of a specific type become adult ones immediately, and differential equations are derived. These things do not happen in real situation case, and in that case a period to become adult should be taken into account and the differential equations in the modified model would become delay differential equations. Just to discuss differential equations in simple forms this adult-assumption is made in this section.

Let $z(t,T)$ denote the number of a particular type of mosquitoes spreading a particular type of disease at a general time t in a particular temperature T in Kelvin scale, in a particular region. Then one has the equations

$$\frac{dz}{dt} = k_T\, z(t,T), \tag{21}$$

where k_T is a nonnegative constant which depends on a particular temperature T, which may also assume value zero. Let us assume as before that $T_1, T_2, T_3, \ldots$ are successive simplified temperatures. Let us assume for simplicity that $z(0,T) = c_0$ so that (21) is equivalent to the following integral equation.

$$z(t,T) = c_0 + k_T \int_0^t z(s,T)\, ds. \tag{22}$$

The corresponding Picard's iteration formula is

$$z_{n+1}(t,T) = c_0 + k_T \int_0^t z_n(s,T)\, ds. \tag{23}$$

Now, a modified synchronization method corresponding to these formulas (23) and corresponding to temperature variations can be used. Let $u(t,T)$ denote the number of humans in the regions who are not affected by the disease spread by mosquitoes at time t in a particular temperature T. Then one has the equation

$$\frac{du}{dt} = u(t,T) - l_T\, u(t,T)\, z(t,T), \tag{24}$$

where l_T is a nonnegative constant which may be interpreted as the possibility to get infected when a mosquito bites a human who has not been infected. Let us assume that $u(0,T) = d_0$ so that (24) is equivalent to the following integral equation.

$$u(t,T) = d_0 + \int_0^t \left(u(s,T) - l_T\, u(s,T)\, z(s,T) \right) ds. \tag{25}$$

The corresponding Picard's iteration formula is

$$u_{n+1}\left(t,T\right)=d_0+\int_0^t\left(u_n\left(s,T\right)-l_T\,u_n\left(s,T\right)z_n\left(s,T\right)\right)ds\,. \tag{26}$$

The values for $z_n\left(s,T\right)$ in (26) should be obtained from (23). Thus the population of mosquitoes and the population of uninfected persons are interrelated and they do depend on temperature of environment, and thereby they do depend on climate changes. Moreover, synchronization should be applied simultaneously to the equations (23) and (26).

It should be observed that theory for everything begins with simplifications, and complications are done by observing factors involved in a system and including them one by one. For example, let us consider the previous WBCs-FBs model. It was observed already that there is a necessity for changes in the equations. Let us assume again that $x\left(t\right)$ denote the population of FBs in a person, and $y\left(t\right)$ denote the population of WBCs in that person at a general time. Then, the following differential equations can be formulated.

$$\frac{dx}{dt}=a\,x\left(t\right)-b\,x\left(t\right)y\left(t\right); \tag{27}$$

$$\frac{dy}{dt}=c-d\,x\left(t\right)y\left(t\right), \tag{28}$$

where a,b,c,d are positive constants. The explanations are the followings. The term $a\,x\left(t\right)$ corresponds to the increase in change corresponding to the present population $x\left(t\right)$ of FBs. The term $-b\,x\left(t\right)y\left(t\right)$ corresponds to the decrease in change corresponding to the present encounters $x\left(t\right)y\left(t\right)$. The term c corresponds to the constant increase in population $y\left(t\right)$ in view of constant production of WBCs from stem cells. The term $-d\,x\left(t\right)y\left(t\right)$ corresponds to the to the decrease in change corresponding to the present encounters $x\left(t\right)y\left(t\right)$. Refinements should always be done based on our experiences. This is applicable even for the model (21) and (24) for mosquito-infected person model corresponding to climate changes.

Climate changes due to pollution may increase vector borne diseases in some regions and decreases in some regions. So, spreading nature of tropical diseases may depend on nature of regions for all continuously existing climate changes. Models should always be refined. Synchronization methods are always being promising methods to understand nature of models.

ANEMIA AND MALARIA

As it was mentioned above, one should always improve mathematical models by considering additional factors. Let $x(t)$ denote the population of malaria parasites (which are considered as FBs) in a human body at a general time t. Let $y(t)$ denote the population of WBCs in the human body at a general time t. Let $w(t)$ denote the population (in terms of counting) of healthy red blood cells in the human body at a general time t. Here, healthy red blood cells are the ones which can carry oxygen using hemoglobin. Moreover, the healthy red blood cells are the cells which host place for malaria parasites. Once a red blood cell gets/hosts a parasite, then it is called a parasitized red blood cell, Ekvall, H. (2003). So, there is a theory provided in the articles, Allen, S. J. et al., (1997), Mockenhaupt, F. P. et al., (2004), that if the body does not have sufficiently many healthy red blood cells then malaria parasites cannot expand their population and hence there would be a natural immunity. That is, anemia helps to get protection against malaria. Equation for malaria parasites is the following.

$$\frac{dx}{dt} = ax(t) + bw(t) - cx(t)y(t). \tag{29}$$

Here a, b, c are positive constants. The term $ax(t)$ corresponds to the increase due to existing population of malaria parasites. The term $bw(t)$ corresponds to the increase due to existing population of healthy red blood cells. The term $-cx(t)y(t)$ corresponds to the decrease due to encounters of malaria parasites and WBCs. Equation for WBCs is the following usual one.

$$\frac{dy}{dt} = d - ex(t)y(t). \tag{30}$$

Here d, e are positive constants. The term d corresponds to the constant increase due to constant production of WBCs from stem cells. The term $-ex(t)y(t)$ corresponds to the decrease due to encounters of malaria parasites and WBCs. Equation for healthy red blood cells is the following one.

$$\frac{dw}{dt} = f - gx(t)w(t). \tag{31}$$

Here f, g are positive constants. The term f corresponds to the constant increase due to constant production of red blood cells from stem cells. The term $-gx(t)w(t)$ corresponds to the decrease due to encounters of malaria parasites and red blood cells. Let $x(0) = a_0$, $y(0) = b_0$ and $z(0) = c_0$. Then the corresponding Picard's iteration formulas are the followings.

$$x_{n+1}(t) = a_0 + \int_0^t \left[ax_n(s) + bw_n(s) - cx_n(s)y_n(s)\right]ds,$$

$$y_{n+1}(t) = b_0 + \int_0^t \left[d - ex_n(s)y_n(s)\right]ds,$$

$$w_{n+1}(t) = c_0 + \int_0^t \left[f - gx_n(s)w_n(s)\right]ds.$$

One may also include population of mosquitoes in a particular temperature along with these equations and synchronization methods can be applied to find relations when climate changes are also considered.

There are articles like Muriuki, J. M. et al., (2019), which claim that anemia need not help to control malarial parasites. It is simple to guess this possibility in view the equation (29), because the term $bw(t)$ may not have influence over $x(t)$, when b is nearly equal to zero. The constants involved in this model may be different and different depending on different and different particular situations. Depending only on the constants which can be observed by means of experiments, the conclusions can be drawn from the equations. Thus, exact relations can be obtained only from the equations and constants.

FUTURE RESEARCH DIRECTIONS

Models may be improved further by using experimental data. Based on the improvement of the models, all mathematical network algorithms may be modified. More specifically, synchronization methods should be given priority whenever there are possibilities to introduce them.

CONCLUSION

There are very few articles like Maire, N. et al., (2006), for modeling in natural immunity, which is the most wanted one for all in our world. For practical purposes, it is being difficult to achieve natural immunity, even if one has meditation, yoga, exercises, good nutrition, fresh air, clean water etc. The exact problem in getting natural immunity lies in necessity of accommodating (or exposing to) natural weather conditions, natural working conditions and natural relaxations in our life, by avoiding air conditioners (air coolers/air heaters), cars, work addictions, net addictions etc. It seems to be impossible for many people to accommodate certain difficulties. For these people, models have to be designed to control diseases.

One has to design models for analysis of tropical diseases to understand certain theoretical facts. Equilibrium points should be found to control diseases, and medicines should be administered towards these equilibrium points. One should begin from simple models, and then models should be complicated to include all important parameters. Forms of solutions of equations involved in models should be found, and they can be used for curve fitting experimental data. It is advised to apply synchronization mathematical methods to solve equations in dynamic models. Many confronting experimental evidences regarding tropical diseases can be explained by means inferences derived from models. So, mathematical modeling should be used for completion in medical analysis. World Health Organization always aims for better health for everyone. Let us also be towards this aim. Let us also be towards natural immunity for healthier future, although this chapter did not deal with this topic directly.

ACKNOWLEDGMENT

This Book Chapter has been written with the joint financial support of RUSA-Phase 2.0 grant sanctioned vide letter No.F 24-51/2014-U, Policy (TN Multi-Gen), Dept. of Edn. Govt. of India, Dt. 09.10.2018, UGC-SAP (DRS-I) vide letter No.F.510/8/

DRS-I/2016(SAP-I) Dt. 23.08.2016 and DST (FIST - level I) 657876570 vide letter No.SR/FIST/MS-I/2018-17 Dt. 20.12.2018.

REFERENCES

Allen, S. J., O'donnell, A., Alexander, N. D. E., Alpers, M. P., Peto, T. E. A., Clegg, J. B., & Weatherall, D. J. (1997). α+-Thalassemia protects children against disease caused by other infections as well as malaria. *Proceedings of the National Academy of Sciences of the United States of America, 94*(26), 14736–14741. doi:10.1073/pnas.94.26.14736 PMID:9405682

Anderson, R. M. (1988). The role of mathematical models in the study of HIV transmission and the epidemiology of AIDS. *Journal of Acquired Immune Deficiency Syndromes, 1*(3), 241–256. PMID:3216309

Bañuelos, S., Martinez, M. V., Mitchell, C., & Prieto-Langarica, A. (2019). Using mathematical modelling to investigate the effect of the sexual behaviour of asymptomatic individuals and vector control measures on Zika. *Letters in Biomathematics, 6*(1), 1–19. doi:10.1080/23737867.2019.1624631

Bernoulli, D., & Blower, S. (2004). An attempt at a new analysis of the mortality caused by smallpox and of the advantages of inoculation to prevent it. *Reviews in Medical Virology, 14*(5), 275–288. doi:10.1002/rmv.443 PMID:15334536

Chitnis, N., Schapira, A., Schindler, C., Penny, M. A., & Smith, T. A. (2018). Mathematical analysis to prioritise strategies for malaria elimination. *Journal of Theoretical Biology, 45*(5), 118–130. doi:10.1016/j.jtbi.2018.07.007 PMID:30006002

Chowell, G., Mizumoto, K., Banda, J. M., Poccia, S., & Perrings, C. (2019). Assessing the potential impact of vector-borne disease transmission following heavy rainfall events: A mathematical framework. *Philosophical Transactions of the Royal Society of London. Series B, Biological Sciences, 374*(7), 1–9. PMID:31056044

Ekvall, H. (2003). Malaria and anemia. *Current Opinion in Hematology, 10*(2), 108–114. doi:10.1097/00062752-200303000-00002 PMID:12579035

Eubank, S., Guclu, H., Kumar, V. A., Marathe, M. V., Srinivasan, A., Toroczkai, Z., & Wang, N. (2004). Modelling disease outbreaks in realistic urban social networks. *Nature, 429*(6), 180–184. doi:10.1038/nature02541 PMID:15141212

Gervas, H. E., Opoku, N. K. D. O., & Ibrahim, S. (2018). Mathematical Modelling of Human African Trypanosomiasis Using Control Measures. *Computational and Mathematical Methods in Medicine*, *2018*(5), 1–13. doi:10.1155/2018/5293568 PMID:30595713

Githeko, A. K., Lindsay, S. W., Confalonieri, U. E., & Patz, J. A. (2000). Climate change and vector-borne diseases: A regional analysis. *Bulletin of the World Health Organization*, *78*(4), 1136–1147. PMID:11019462

Greenwood, M. (1916). The application of mathematics to epidemiology. *Nature*, *97*(4), 243–244. doi:10.1038/097243a0

Hethcote, H. W. (2000). The mathematics of infectious diseases. *SIAM Review*, *42*(4), 599–653. doi:10.1137/S0036144500371907

Hippner, P., Sumner, T., Houben, R. M., Cardenas, V., Vassall, A., Bozzani, F., & White, R. G. (2019). Application of provincial data in mathematical modelling to inform sub-national tuberculosis program decision-making in South Africa. *PLoS One*, *14*(1), 1–11. doi:10.1371/journal.pone.0209320 PMID:30682028

Keeling, M. J., & Eames, K. T. (2005). Networks and epidemic models. *Journal of the Royal Society, Interface*, *2*(4), 295–307. doi:10.1098/rsif.2005.0051 PMID:16849187

Kermack, W. O., & McKendrick, A. G. (1927). A contribution to the mathematical theory of epidemics. *Proceedings of the Royal Society of London. Series A, Containing Papers of a Mathematical and Physical Character*, *115*(772), 700–721. doi:10.1098/rspa.1927.0118

Kermack, W. O., & McKendrick, A. G. (1932). Contributions to the mathematical theory of epidemics. II.—The problem of endemicity. *Proceedings of the Royal Society of London. Series A, Containing Papers of a Mathematical and Physical Character*, *138*(834), 55–83. doi:10.1098/rspa.1932.0171

Kermack, W. O., & McKendrick, A. G. (1933). Contributions to the mathematical theory of epidemics. III.—Further studies of the problem of endemicity. In *Proceedings of the Royal Society of London. Series A, Containing Papers of a Mathematical and Physical Character*, *141*(843), 94–122. doi:10.1098/rspa.1933.0106

Kreyszig, E. (1978). *Introductory functional analysis with applications*. New York: Wiley.

Maire, N., Smith, T., Ross, A., Owusu-Agyei, S., Dietz, K., & Molineaux, L. (2006). A model for natural immunity to asexual blood stages of Plasmodium falciparum malaria in endemic areas. *The American Journal of Tropical Medicine and Hygiene*, *75*(2), 19–31. doi:10.4269/ajtmh.2006.75.19 PMID:16931812

Mockenhaupt, F. P., Ehrhardt, S., Gellert, S., Otchwemah, R. N., Dietz, E., Anemana, S. D., & Bienzle, U. (2004). α+-thalassemia protects African children from severe malaria. *Blood*, *104*(7), 2003–2006. doi:10.1182/blood-2003-11-4090 PMID:15198952

Moorthy, C. G., & Raj, S. I. (2018). Fixed points of sequences of mappings, *Results in Fixed Point Theory and Applications, 2018*(1), 1-10.

Muriuki, J. M., Mentzer, A., Band, G., Gilchrist, J., Carstensen, T., Lule, S., & Cutland, C. (2019). The ferroportin Q248H mutation protects from anemia, but not malaria or bacteremia. *Science Advances*, *5*(9), 1–12. doi:10.1126ciadv.aaw0109 PMID:31517041

Musa, S. S., Zhao, S., Chan, H. S., Jin, Z., & He, D. (2019). A mathematical model to study the 2014–2015 large-scale dengue epidemics in Kaohsiung and Tainan cities in Taiwan, China. *Mathematical Biosciences and Engineering*, *16*(5), 3841–3863. doi:10.3934/mbe.2019190 PMID:31499639

Newfield, T. P. (2016). Mysterious and Mortiferous Clouds: The Climate Cooling and Disease Burden of Late Antiquity. *Late Antique Archaeology*, *12*(1), 89–115. doi:10.1163/22134522-12340068

Rock, K. S., Ndeffo-Mbah, M. L., Castaño, S., Palmer, C., Pandey, A., Atkins, K. E., & Chitnis, N. (2018). Assessing strategies against Gambiense sleeping sickness through mathematical modeling. *Clinical Infectious Diseases*, *66*(4suppl_4), 286–292. doi:10.1093/cid/ciy018 PMID:29860287

Serfling, R. E. (1963). Methods for current statistical analysis of excess pneumonia-influenza deaths. *Public Health Reports*, *78*(6), 494–506. doi:10.2307/4591848 PMID:19316455

Shirley, M. D., & Rushton, S. P. (2005). The impacts of network topology on disease spread. *Ecological Complexity*, *2*(3), 287–299. doi:10.1016/j.ecocom.2005.04.005

Tjaden, N. B., Caminade, C., Beierkuhnlein, C., & Thomas, S. M. (2018). Mosquito-borne diseases: Advances in modelling climate-change impacts. *Trends in Parasitology*, *34*(3), 227–245. doi:10.1016/j.pt.2017.11.006 PMID:29229233

ADDITIONAL READING

Simmons, G. F. (1972). *Differential Equations with applications and historical notes*. New York: McGraw-Hill.

KEY TERMS AND DEFINITIONS

Equilibrium Points: They are points (x(s),y(s)) for solution curves x and y of a system of differential equations with independent variable t, at which x(s) and y(s) are equal.

Fixed Points: They are points x* for a function f from a set X to itself such that x* are members of X, and such that f(x*) coincides with x*.

Heat Rashes: They are rashes which appear due to above normal heat in climate changes.

Mathematical Modeling: Converting real life situations into mathematical concepts and symbols and thereby converting real life problems into mathematical problems.

Network Modeling: It is a mathematical modeling in which the problems or procedures to solve the problems may be described in terms of a network comprising of nodes and paths connecting nodes.

Stability: This is a required behavior of parametric solution curves in neighborhoods of an equilibrium points.

Stem Cells: They are basic cells from which red blood corpuscles and white blood corpuscles are produced.

Synchronization Methods: They are mathematical methods which consider some equation as a prime equation and the methods always adjust solutions for secondary equations as approximate solutions of the prime equation.

Section 3

Climate Change and Anthropogenic Impacts on Some Tropical Neglected Diseases From Two Vulnerable Continents: Africa and Asia

Chapter 5 Strongyloidiasis: Biology, Diagnosis, and Management of a Most Neglected Tropical Disease

Junaid Ahmad Malik
https://orcid.org/0000-0003-4411-2015
Government Degree College Bijbehara, India

ABSTRACT

Strongyloidiasis is a human parasitic ailment brought about by a whiplike nematode worm called Strongyloides stercoralis. Most humans get the infection by coming in contact with contaminated soil whereby the tiny worms penetrate the skin and enter the bloodstream from where it passes through the right side of the heart and lungs to mouth, stomach, and small intestines. Replication inside tainted people enables the disease to persevere for quite a long time. Strongyloides can cause a hyperinfection disorder which causes side effects in numerous organ frameworks, including the central nervous system that can prompt death if untreated. The diagnosis is made by blood and stool tests. This chapter is intended to draw a more precise picture of the global prevalence, diagnosis, and risk factors for S. stercoralis. The chapter also discusses the diagnostic approaches for detecting the infection, the morbidity caused and the recommended management. It further discusses some of the reasons why this infection is so neglected and the consequence of this for the estimated global prevalence.

DOI: 10.4018/978-1-7998-2197-7.ch005

INTRODUCTION

Strongyloidiasis is a human parasitic illness brought about by a nematode called *Strongyloides stercoralis*, or some of the times by *Strongyloides fülleborni* which is a category of helminth. This intestinal worm can cause various side effects in individuals, chiefly skin manifestations, stomach torment, loose bowels and weight reduction, among numerous other explicit and unclear side effects in dispersed illness, and extreme dangerous conditions through hyperinfection. In certain individuals, especially the individuals who require corticosteroids or other immunosuppressive medicine, *Strongyloides* can cause a hyperinfection disorder that may cause death if untreated (Buonfrate et al, 2015; Vazquez-Guillamet et al., 2017). The diagnosis of the disease is done by blood and stool tests. The disease is common in tropical and subtropical areas of the world. *S. stercoralis* disease likewise is endemic in certain parts of Europe, the south eastern United States and Puerto Rico. Clinically, strongyloidiasis is critical in the light of the fact that disease perseveres for a long time in the human host, more often than not as an undetected, asymptomatic condition. In the zones of endemicity and non-endemicity (on account of movement and so on), this drawn out course of disease results in a critical however unsuspected worm load in the tainted populace. Under explicit conditions (e.g., immunosuppression), this undetected worm burden produces broad tissue attack by parasite hatchlings, the supposed hyperinfection disorder. This disorder is a genuine ailment that requires a high record of doubt and forceful, early remedial intercession. Recently there has been a resurgence of intrigue in *Strongyloides stercoralis*. The across the board utilization of immunosuppressive and chemotherapeutic medications may have brought about a genuine increment in the occurrence of scattered strongyloidiasis; or, maybe, as parasitic diseases are lessening in industrialized nations, the sporadic appearance of a serious, regularly lethal disease brought about by a worm is considered more deserving of detailing. Whatever the reason, nearly 300 instances of pioneering *S. stercoralis* contamination in immune-compromised patients have been distributed over the most recent 20 years, and some thorough surveys have outlined the clinical attributes of these patients (Igra-Siegman et al., 1981; Longworth et al., 1986; Scowden et al. 1978).

Organically, *S. stercoralis* is unusual among human helminths, which don't increase inside the host and the worm is fit for reproducing and re-erecting inside the human host. The typical autoinfection cycle includes the arrival of parasite hatchlings by grown-up worms in the gastrointestinal tract, and rather than going to the outside condition in dung, they change into infective hatchlings and infiltrate the gut divider or peri-anal skin of the same person. Increasing rapidity of this autoinfection cycle results in the hyperinfection disorder with broad tissue intrusion by hatchlings. The existence cycle of *S. stercoralis* is further convoluted by its capacity to seek after either parasitic or free-living examples of presence. These natural attributes assume a noteworthy job in forming the result of the hostparasite communication. *S. stercoralis* larvae were first described in 1876 in stools of French troops with severe diarrhoea who were returning from Indochina (Machicado et al., 2012; Politis et al., 2017).

This invigorated enthusiasm for the clinical and essential parts of strongyloidiasis has been paralleled in epidemiologic research. In the United States alone, five examinations intended to evaluate the pervasiveness of *S. stercoralis* disease in explicit territories or high-hazard crowds have been distributed since 1980. Furthermore, a few agents have displayed enormous arrangement of patients to describe the clinical viewpoints and the hazard components of strongyloidiasis. While not carefully epidemiologic, the last examinations give valuable signs about the significance of this parasitosis in the territories where they were led. The subject of the commonness of strongyloidiasis has down to earth intrigue. Most lethal scattered contaminations that happen in immunosuppressed patients are preventable by early discovery and treatment. The analysis can be cultivated with a humble venture of research centre assets. However, the parasitological screening of all possibility for immuno-suppressive or antineoplastic treatment might be a purposeless exercise in regions where *S. stercoralis* is exceedingly remarkable. Then again, where this parasitosis is present, screening may spare lives and expensive intercessions. Data with respect to the commonness of strongyloidiasis in various pieces of the world is fragmentary, isn't promptly available, and has not been basically assessed. The reason for this paper is to inspect the information from countless predominance considers and to figure rules for the foundation of early-location programs.

The amount and nature of data produced by various examinations mirror the heterogeneity of the sources. A few papers report on painstakingly arranged planned investigations and give fine methodological subtleties. In different articles (for instance, those detailing on all parasites found in the feces tests of a given populace), information on the pervasiveness of strongyloidiasis might be consigned to a table with no remark. We have endeavoured, at whatever point the first source gave adequate foundation data, to evaluate fundamentally for the peruser the legitimacy of the data given by each paper. The troubles of social occasion important information on the study of disease transmission of intestinal parasites have as of late been checked on by Crompton and Tulley (1987). The accumulation of stools for parasitological examination is a strategy that regularly helps not experience out of mobile patients. In these cases consistence is identified with the degree of inspiration, and a patient will constantly gather at least one specimen just on the off chance that the individual in question has been persuaded of the significance of the test. In this way, unmistakably the accumulation of various fecal specimens from an enormous number of sound people in a network for the sole reason of assessing the predominance of a parasite might be incredibly difficult.

Strongyloidiasis

The worldwide pervasiveness of strongyloidiasis is assessed to be 100 million cases (Nutman & Krolewiecki, 2019). This is a general evaluation, and the commonness of this disease changes extensively in various territories. In the US, the most noteworthy commonness rates (4%) are found in eastern Kentucky and provincial Tennessee (Mejia & Nutman, 2012). There are three groups of people for whom strongyloidiasis speaks to a critical issue: occupants of mental establishments (commonness rates 1.7%-40%), veterans who were detainees of war in territories of endemicity during World War II (predominance rates 0.5%-37%), and outcasts (refugees) and outsiders to the United States and Europe (commonness rates 0.6%-38%) (Genta, 1989). Data on the study of disease transmission of the hyperinfection disorder of strongyloidiasis is less precise. Certain patient groups seem, by all accounts, to be progressively helpless to this disorder, incorporating those with malignancies (particularly hematologic), immune system maladies, furthermore, hunger (Forrer et al., 2017). The disorder additionally is as often as possible seen in patients getting immunosuppressive treatment, including corticosteroids. The relationship of immunosuppression due to HIV or human T lymphotropic infection disease and hyperinfection with *S. stercoralis* has been seen in just a couple of conditions.

General reviews achieved in different nations show that *S. stercoralis* is exceptionally pervasive in parts of Latin America. Numerous investigations have been performed in Brazil, with detailed predominance rates between 15% and 82% (Carvalho, 1978). Over the span of a broad examination of more than 3,000 schoolchildren led in the mid 1970s in the city of Salvador, Bahia, incredible 62% were observed to be tainted (Faria, 1972; Khieu et al., 2013). However another surprise is the predominance of 32% proclaimed in 1,574 working class grown-ups living in the city of Guayaquil, Ecuador (Alvarez, 1978). As per the author, *S. stercoralis* hatchlings were available in the pee of 30% of patients with positive stools. This is an extraordinarily uncommon finding, and one must think about how conceivable it is that nonparasitic material was confused with hatchlings. Along these lines, the legitimacy of the whole investigation shows up sketchy. In Costa Rica, various overviews of the pervasiveness of helminths in feces tests have been directed in the previous 30 years, and *S. stercoralis* has been found at rates between 1.1% and 16.5%, contingent upon the locale (Arroyo et al., 1985). An overview of intestinal parasitic diseases in an sample of 203 youngsters aged 3-5 years from a semi urban and provincial community in Chiriqui, Panama, was performed during 1983-1984 (Holland et al., 1987). *S. stercoralis* was found in eight of the youngsters when a solitary feces test was analysed by the formalin concentration method. Be that as it may, one-fourth of the kids in this survey were chosen since they were known to harbour *Ascaris lumbricoides*, and in this way the test was not by any means free of determination predispositions. The greater part of these studies is review investigations of stool examination records, and the data provided does not generally enable one to perceive whether the level of positives alludes to the quantity of fecal samples found to contain the parasite or to the number of people observed to be contaminated. In Europe indigenous strongyloidiasis is existing as far north as Nottingham, England (Sprott et al., 1987). Clinical investigates arrangement of patients who obtained the parasite locally have been distributed in France (Junod, 1987), Switzerland (Berthoud & Berthoud, 1975), Portugal (Viera, 1979), Italy (Scaglia et al., 1984) and Yugoslavia (Breitenfeld, 1984). Individual privately procured instances of dispersed sickness have happened in Spain (Capellet al., 1978; Ruiz et al., 1987) and Greece (Giannoulis et al., 1986). An ongoing medical clinic-based review of 4,203 patients in northern Italy uncovered a pervasiveness of 3.0% (Genta et al., 1988). This investigation was intended to assess the nearness of strongyloidiasis in a region where this parasitosis had been known to be normal (Scaglia et al., 1984). Notwithstanding the routine examination of three crisp feces tests, the Baermann technique for larval fixation, or baermanization (Lima & Delgado, 1961), was utilized when strongyloidiasis was emphatically suspected clinically. A pervasiveness of 26070 has been accounted for from the Congo (Junod, 1987) and of up to 48070 from the Central African Republic (Brumpt et al., 1972). In Nigeria the pervasiveness of *S. stercoralis* shifts somewhere in the range of 0.2% and 11.7%, depending on the zone and the populace surveyed (Ejezie & Otigbuo, 1987). Available epidemiologic information from Asia is limited. At the Hacettepe clinics in Ankara, Turkey, *S. stercoralis* was found in just 7 of 26,097 stool tests analysed from 1974 to 1979 (Igra-Siegman et al., 1981; Scowden et al., 1978). The commonness of strongyloidiasis on the Indian subcontinent is obscure. As of late, a focal point of obvious high commonness in the Punjab state has been reported, and a pilot epidemiologic examination is under way.

Nutman and Krolewiecki (2019) reported that, from a general wellbeing point of view, the assessed size of the populace influenced who are in danger and its relationship to neediness and absence of sufficient water and sanitation, put strongyloidiasis unequivocally in the neglected tropical disease (NTD) camp, in spite of the fact that it isn't right now amalgamated in the World Health Organization's (WHO) technique against STH (WHO, 2017) notwithstanding the developing acknowledgment of strongyloidiasis as an illness of incredible general wellbeing centrality (Krolewiecki et al., 2013; Olsen et al., 2009).

MORPHOLOGY OF *STRONGYLOIDES STERCORALIS*

S. stercoralis has a parasitic life cycle as well as a free-living phase. Key features of these forms are as;

Parasitic Female Adult Worm

This stage is a minute, slim, practically transparent worm a little more than 2 mm long. The mouth is hexagonal fit as a fiddle and encompassed by six papillae. The body is constricted anteriorly and contains the long tube shaped oesophagus (in some cases called the pharynx). The pharynx is prevailing by the intestine, a long slight cylinder, that finishes in a short rectum which opens at the rear-end near the tip of the tail (Figure 1). The pharynx is encompassed by the nerve ring at around one quarter of the route down its length. There are two longitudinal excretory trenches running the length of the worm, one in every parallel harmony, and associated by a transverse pipe to one another and a solitary excretory cell simply behind the nerve ring. The vulva is situated in the mid-ventral line in the back third of the body. Paired uteri broaden anteriorly and posteriorly from an extremely short vagina. The uteri contain few eggs adjusted in single record and possess a large portion of the body of the worm in this locale. Each uterus opens into an oviduct which prompts thin walled ovaries.

Free-living Adult Female Worm

This little worm looks like the free-living nematodes of the sort *Rhabditis*. The fingernail skin (cuticle) is thin, transparent and has fine ridges. There are three parts of the oesophagus - the foremost, tube shaped procorpus, the tight isthmus, and the rouded back bulb (Figure 1). This thus prompts the intestine then rectum which opens at the rear-end close to the tail. The reproductive system is like that found in the parasitic female with the exception of that every uterus contains various eggs.

Free-living Adult Male Worm

The littler males have a strongly pointed tail that curves anteriorly and gives the worm a "J-shape". The gut is like that found in the female worm. The reproductive system is a transparent straight cylinder or a simple tube. At the foremost end is a blind ended testis that converges without clear outline first into the vas deferens then into the seminal vesicle. These organs contain spermatogonia, spermatocytes and sperms and open into the cloaca. The cloaca is encompassed by a couple of copulatory spicules that are embedded into the female during copulation. The spicules are guided in their expulsion by a chitinoid structure in the dorsal mass of the cloaca called the gubernaculum. There are six sets of caudal papillae and a solitary midventral precloacal papilla (Figure 1).

Figure 1. The picture depicts each stage of development of Strongyloides. The drawing on the far left depicts an adult female parasitic worm. The middle one depicts a rhabditiform larva; notice the obvious rhabditiod esophagus towards the top end, with a club-shaped anterior (towards the top) portion and short buccal cavity, a constriction in the middle, and a bulb posteriorly. The drawing on the far right represents a filariform stage larva; notice that it is longer and skinnier than the rhabditiform stage
(Source: Little, 1966).

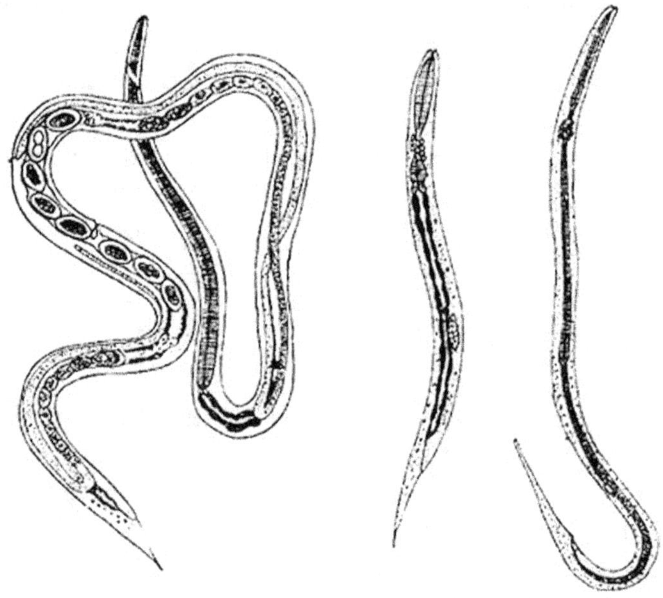

LIFE CYCLE AND INFECTION TO HUMANS

The verifiable perceptions of purposeful trial infection of people with *S. stercoralis* are assessed as are those of infections with different *Strongyloides* species. Experimental human infections have given exceptional chances to think about the connections between specific parasites and their characteristic human hosts. In any case, since a great part of the essential science of *S. stercoralis* was comprehended by the mid-1930s, hardly any issues of major physiologic significance were left to be tended to by purposeful human infections. As *S. stercoralis* effectively infects primates, hounds, and other little creatures, exemplary early examinations, for example, those by Fiillebom in 1926 and Sandground in 1926 and 1952, used these incredible and promptly accessible creature models to set up what includes the mass of our present comprehension of the essential science of the creature/organism.

The infection of *S. stercoralis* is instigated by the penetration of infective falariform larvae into the skin (Figure 2). The infection with the early stage of the *S. stercoralis* in the human host is usually having no reasonable symptoms. Hatchling currens is a more frequent cutaneous indication that happens in relationship with autoinfection. It is an unfavourably susceptible response to filariform hatchlings that relocate in the skin at a moderately quick rate and leave irritated urticarial convoluted tracks. The condition is generally identified over the lower abdomen, buttocks or thighs. Hatchling currens may keep going for a couple of days however for the mostly recurs over delayed periods (Vazquez-Guillamet et al., 2017).

Investigations of tentatively tainted people have been blocked throughout the years by the presence of the strongyloides hyperinfection disorder also as the parasite's trademark unmanageability to medicate treatment. With the goal that the test human diseases talked about in this audit are put in setting, ebb and flow ideas of malady pathobiology that were grown to a great extent because of the previously mentioned organism models will be quickly audited. Strongyloides organisms have a mind boggling life cycle that has three separate formative stages: grown-up worms, rhabditiform hatchlings, and filariform (irresistible) hatchlings. Three option clinically significant methods of replication exist. In the inside sexual cycle, the grown-up female dwells in the small digestive tract what's more, lays few eggs that incubate very quickly. The subsequent rhabditiform hatchlings are passed in the stool and develop in the dirt to infective filariform hatchlings; the infective hatchlings infiltrate the skin of another host. After relocation through the lung, the hatchlings creep over the glottis and are gulped; their advancement is finished in the jejunum. Then again, in the free-living cycle, rhabditiform hatchlings may rather form in the dirt into explicitly dynamic grown-ups, in this way making an ecological supply of grown-up worms (Khieu et al., 2013; Vazquez-Guillamet et al., 2017). Each age of this outside sexual cycle incorporates another accomplice of irresistible filariform hatchlings prepared to reappear the parasitic cycle in people (Figure 2).

At last, in the autoinfective cycle, irresistible filariform hatchlings create from rhabditiform hatchlings while still in the digestive tract. Infiltration of the colon or the butt-centric skin by filariform hatchlings permits reinfection of a similar host. Of significance, the autoinfective cycles, which typically result in average constant contamination in the hosts, ought to be recognized from the hyperinfective cycles. Hyperinfection results when autoinfective cycles are uncontrolled by the host's resistant reaction and enormous, much of the time deadly, dispersal of hatchlings to various far off organs follows.

Figure 2. Pattern of infection (cycle) of Strongyloides stercoralis
(Source: Roger & Gonzalez, 2017)

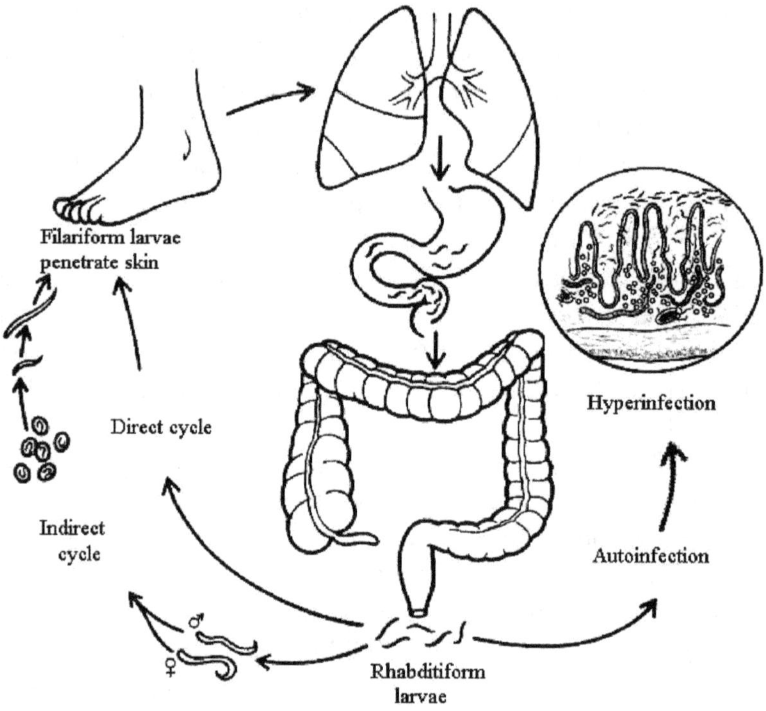

CHRONIC INFECTION AND IMMUNOLOGY

A few people mount an in part viable invulnerable reaction. They can't destroy the infection yet they do contain the force and vigour of disease. Various proteins superficially or in the excretory/secretory results of *S. stercoralis* infective hatchlings have been radiolabelled; some of these were immunogenic and encouraged human serum antibodies or invigorated IgE-interceded arrival of histamine from basophils of infected people (Brindley et al., 1988). Sato et al. (1990) discovered 33-39 groups on sodium dodecylsulphate-polyacrylamide gel electrophoresis (SDS-PAGE) examination of *S. stercoralis* entire body larval concentrate. On consequent immunoblot examination, there was impressive changeability among sera from tainted patients with the quantity of responsive groups changing between one what's more, more than 18. The most noticeable groups were 26, 41, 66 and 99 kDa in estimate. In another examination, three proteins of *S. stercoralis* filariform hatchlings of 28, 31 and 41 kDa in size were perceived on Western smudges by serum IgG antibodies from tainted patients (Conway et al., 1993; Lindo et al., 1994).

Further investigations demonstrated that these antigens were solvent in both watery and cleanser separates, did not share epitopes between them, were not found in *S. cebus* or *S. rarti*, and responded with antibodies dominatingly in the IgG_1 and IgG_4 yet not IgG_2 and IgG_3 parts of human serum (Conway et al., 1994). Likewise, Genta and Lillibridge (1989) had prior noticed that IgG_4 antibodies were most conspicuous in human strongyloidiasis. In an animal model in which Balb/cByJ mice were vaccinated with infective *S. stercoralis* hatchlings, 97% of parasites in a test disease were killed inside 24 hours and this impact was reliant on the nearness of eosinophils (Rotman et al., 1994; Spinicci et al., 2017). Whenever S. stercorulis infective hatchlings were set in dissemination chambers built with films containing pores changing from 0.1 to 2.0 pm in size and were then embedded in recently inoculated mice, hatchlings were possibly executed if there was contact among leucocytes and parasites (Abraham et al., 1995). Concentrates in vitro have demonstrated that living *S. stercoralis* filariform hatchlings actuate the supplement framework by both old style and exchange pathways and that initiated supplement, without explicit immune response, encourages grip of fringe blood mononuclear and polymorphonuclear cells to hatchlings. Supplement segments C1q, C3, C4, C8 and properdin can be found on the larval surface and hatchlings covered with white cells lose their motility (Messias et al., 1994). The supplement framework in relationship with effector cells may along these lines play a significant non-explicit job in the first line of barrier against relocating parasites.

Lymphocyte movement was evaluated in 64 patients with human strongyloidiasis by Sato and Shiroma (1989) in Okinawa. They found a critical increment of $CD4^+$ and $OKIa1^+$ cells, a relative reduction in $CD8^+$ cells, expanded unconstrained mitogenesis and interleukin 2 generation, however a brought lymphoproliferative reaction down to mitogens. Considerably increasingly baffling are the methods by which hatchlings that are coursing in the tissues and circulatory system can avoid the cell and humoral barrier components in these patients. Albeit inconspicuous increments in pole cell numbers happen in the digestive tract of Erythrocebus patas monkeys contaminated with *S. stercoralis* (Barrett et al., 1988), parasites in the human little intestinal mucosa inspire little in the method for an incendiary reaction in this manner suggesting that here and there they can avoid or smother invulnerable instruments. In patients with endless diseases, a parity is come to with neither the parasite nor the patient picking up the high ground. Albeit extreme diseases were depicted sporadically in the before portions of this century, it is just over the most recent 20 years that the condition has become progressively perceived and its relationship with immunosuppression, particularly impedance of cell-intervened resistance, increased in value. Hidden sicknesses in patients with extreme, entangled strongyloidiasis incorporate lymphoma, intense and interminable leukemia, carcinoma, endless glomerulonephritis, nephrotic disorder and renal transplantation, interminable lung illness, fundamental lupus erythematosus, idiopathic thrombocytopenic purpura, polymyositis, hypercalcaemia, eye illnesses, skin sicknesses including bums and sickness, perpetual liquor addiction, lack of healthy sustenance and retroviral diseases. As far as detailed cases, by a wide margin the most well-known basic condition has been in patients who have experienced renal transplantation (DeVault et al., 1990; Khieu et al., 2013).

CLINICAL FEATURES AND SYMPTOMS

Numerous patients incessantly infected with *S. stercoralis* are asymptomatic. Others have an assortment of indications, generally referable to the skin or gastrointestinal framework. The real focuses of disease are the gut, lungs and focal sensory system. The mortality in patients with immense diseases is high, reflecting the parasitic disease, yet in addition the hidden condition which inclines to dispersal. Gastrointestinal indications incorporate heartburn, cramping lower stomach torments, irregular or tireless looseness of the bowels, pruritus ani, and in some cases weight reduction. These patients might be misdiagnosed as having fractious gut disorder. Patients may gripe of fart, queasiness, stomach distension, visit noxious stools, and weight reduction or puffiness of the face and lower legs. A few patients have highlights of intestinal impediment with fever, tachycardia, hypotension and stomach distension, delicacy and missing gut sounds characteristic of incapacitated ileus (Bannon et al., 1995; Forrer et al., 2017). Different introductions incorporate necrotizing jejunitis, arteriomesenteric impediment (Lee and Terry, 1989), small bowl infarction (Kennedy et al., 1989), papillary stenosis with biliary impediment (Delarocque-Astagneau et al.,1994; Khieu et al., 2013).

Patients usually whine of an irritative or painful cough and brevity of breath, often concomitant with wheezing. Examination generally uncovers dissipated pops and wheezes. In extraordinary cases, respiratory disappointment creates and may require intubation and ventilation. To be sure, one case has been depicted in which lethal grown-up respiratory misery disorder seemed to create after effective treatment of the parasitic contamination (Thompson and Berger, 1991).

Intrusion of the focal sensory system by moving hatchlings is frequently accompanied by auxiliary bacterial contamination. This may bring about meningitis or brain abscess causing any mix of fever, migraine, queasiness, heaving, neck firmness, or seizures or trance like state. Lumbar cut may uncover proof of bacterial meningitis with expanded neutrophils and protein fixation yet a decreased glucose level in the cerebrospinal fluid (Satoh and Kokaze, 2004; Spinicci et al., 2017).

DIAGONOSIS AND TESTS

The exemplary set of three of the runs diarrhea, stomach agony and urticaria is suggestive of a finding of strongyloidiasis and intimations are given by the disclosure of an eosinophilia or suggestive radiological discoveries and upheld by serological tests. Parasites are normally found in defecation however once in a while they are seen in other body liquids or tissue tests. Distinguishing proof is most precise when grown-up as well as larval stages are accessible for examination however this is infrequently conceivable with clinical examples.

Stool Examination

The parasitological determination is generally made after examination of excrement. Most clinical labs will distinguish *S. strongyloides* just upon observation of rhabditiform hatchlings in the faeces and will normally be right. It has been proposed that rhabditiform hatchlings of *S. stercorulis* lash out like a whip though those of hookworm coast crookedly like a snake (Gustavo, 1993). Rehashed examinations of stool examples improve the odds of discovering parasites (Machicado et al., 2012; Nielsen and Mojon, 1987; Woodring, 1994). In the Baermann concentration strategy, stool is set on coarse texture or paper overlying a work screen in a pipe that is loaded up with warm water and associated with a clipped cylinder; after brooding, hatchlings move into the water and are gathered by sedimentation in the water through centrifugation. In spite of its unrivalled viability in diagnosing contaminations in various examinations, this strategy is bulky and relentless (Requena et al., 2002; Siddiqui and Berk, 2001) and other methods, for example, the Harada-Mori filter paper culture or nutrient agar plate cultures are likewise techniques with an affectability higher than direct spreads/smears (however require longer brooding times). Affectability has improved with bigger quantities of successively collected samples, achieving practically 100% when 7 faeces tests were considered (Dreyer et al., 1996; Siddiqui and Berk, 2001).

The formalin-ether concentration method, or some alteration of it, in which about 0.5 g of stool are inspected, is ordinarily utilized. One slight disservice of the formalin-ether technique, in correlation with an immediate smear, is that hatchlings are slaughtered by the formalin what's more, are in this way not all that effectively distinguished by the microscopist as are motile hatchlings. The more extended and progressively thin filariform hatchlings may likewise be available in the specimen. By chance, either rhabditiform or filariform hatchlings may likewise be available in the sputum; the previous sort of hatchling demonstrates the nearness of grown-up female worms in the lungs, though filariform hatchlings may basically be traveling the lungs.

Serological Test

Serological diagonosis utilizing an enzyme-linked immunosorbent assay is presently accessible in numerous centres. Affectability and particularity may both be improved by pre-hatching of serum with *Onchocerca gutturosa* antigens (Conway et al., 1993; Lindo et al., 1994; Spinicci et al., 2017). There has been significant disarray about the pertinence of sero-diagnosis at the point when applied to people and to populaces. Genta (1988) asserted that the ELISA was 88% touchy, 99% explicit, and had positive and negative prescient estimations of 97% and 95% respectively, however was at that point completely scrutinized for utilizing erroneous insights, and furthermore for inability to consider the commonness in the populace (Lo and Kajioka, 1989; Machicado et al., 2012). The significant incentive in serology is the arrangement of a screening test which, if positive, can invigorate further scans for the parasite.

Radio-Imaging

Most patients have typical chest radiographs despite the fact that the intermittent patient has penetrates predictable with Loeffler's disorder. In patients with enormous strongyloidiasis, there might be broad, diffuse aspiratory opacities (Woodring et al., 1994). In patients with extreme intestinal diseases, circles of widened small bowel might be seen on a plain stomach radiograph. Most patients have typical barium feast examinations however in overwhelming diseases, duodenal dilatation or duodenal stricture with ulceration have been depicted; the pipestem appearance may switch after effective treatment (Medina et al., 1992).

Endoscopy

The duodenum looks typical in most of contaminated patients with uncomplicated sickness in spite of the fact that there have been intermittent reports of gastritis or duodenitis. In patients with serious infection, straightening of the duodenal folds, mucosal oedema and ulcerations might be seen (Chen et al., 1994). Sigmoidoscopy and colonoscopy are commonly ordinary however once more there have been periodic reports of ulceration. In each of these irregular cases, it is hard to be sure that the anomalies watched are in reality due to Strongyloides contamination.

TREATMENT AND MANAGEMENT

Since the parasites are frequently hard to distinguish in any case, the issue is aggravated habitually by vulnerability in deciding if they have all been killed. *S. stercoralis* is moderately impervious to anthelmintics and most consideration has concentrated on benzimidazole and ivermectin. The absence of a reference standard for finding is a genuine issue when the adequacy of a treatment is assessed. A constrained number of clinical preliminaries on *S. stercoralis* treatment have been led during the most recent three decades, generally looking at thiabendazole, albendazole, and ivermectin. Thiabendazole and ivermectin have comparable efficiencies, however the vastly improved decency of the last has settled on it the medication of decision overall (Bisoffi et al., 2014). In direct correlations, ivermectin has been demonstrated to be more powerful and preferable endured over thiabendazole (Gann et al., 1994) and more powerful than albendazole (Datry et al., 1994). It is part of the Avermectin family, derived from the actinomycetes *Streptomyces avermectinius* (Molinari et al., 2010). It causes loss of motion of nematodes through the deluge of chloride particles crosswise over cell films (Ottesen and Campbell 1994). It has a half-life of 18 hours and is basically discharged by defecation. The most generally announced reactions are gentle, including migraine, dazedness, muscle agony, queasiness or the runs. It is directed orally, and as a result of the comparably modest quantity of medication necessary, the Ivermectin tablets are small and easy to swallow. This is a bit of leeway when treating youngsters. In a few case-reports with respect to the viability and wellbeing a parenteral or subcutaneous organization of a veterinary detailing of Ivermectin was directed. This treatment was employed in experimental conditions which were hard to ace utilizing oral substances (Chiodini et al., 2000). A number of studies have demonstrated the pre-eminence of Ivermectin over alternative medicines in terms of effectiveness and side-effects (Adenusi, 2003; Ordonez & Angulo, 2004). Ivermectin is as of now the medication of decision for all sorts of medicines. The routine of 200 mg/kg/d for 2 days remains the treatment for uncomplicated *S. stercoralis* contaminations as it targets both adults and larvae (Mejia & Nutman, 2012; Henriquez et al., 2016).

Thiabendazole is a benzimidazole constraining the mitochondrial helminth-specific enzyme fumarate reductase. The medicine cannot be well-thought-out first choice to any further extent and should only be used if Ivermectin is not available (Zaha et al., 2000). The disappointment rate of thiabendazole treatment is still near 30% (Grove, 1982), and progressively successful medications are currently not accessible. Ivermectin has been observed to be exceedingly viable (>90% in most preliminaries), however the general absence of affectability of fecal determination is liable to have caused an overestimation of the medication adequacy (Requena et al., 2014). Thiabendazole (25 mg/kg/d) for 3 days is an elective treatment, yet drug accessibility and GI unfavorable impacts have constrained the utilization of this medication as ivermectin turned out to be all the more generally accessible. Albendazole at 400 mg two times per day for 3 to 7 days has been demonstrated to be less viable than ivermectin for the treatment of uncomplicated *S. stercoralis* (Henriquez et al., 2016, Suputtamongkol et al., 2011).

Bringing issues to light about the sickness among populaces in danger and medicinal services experts is emphatically suggested (Boggild et al., 2016). During a *S. stercoralis* and *Trypanosoma cruzi* screening campaign, performed in 2016 among Latin-American settlers in Alicante, Spain, it was discovered that 92.2% of members (119/129) had never known about strongyloidiasis, including none of the ten members who had a positive *S. stercoralis* serology (individual correspondence) (Molinari et al., 2010; Woodring et al., 1994). Also, a poll around five Neglected Tropical Diseases finished by scholars from Madrid in their last year of Medicine, uncovered that under 18% of the understudies (18/103) 'passed the test' on strongyloidiasis, this being one of the most stressing outcomes (Roger and Gonzalez, 2017). We reason that screening for strongyloidiasis ought to be obligatory for HIV patients, as well with respect to both transplant beneficiaries and contributors originating from endemic regions. Contamination ought to additionally be discounted in those determined to have HTLV-1 contamination, preferably before the beginning of steroid treatment. Obviously sickness results improve when clinicians know about the contamination and ivermectin supply is accessible for patients who require it.

PREVENTION

There seems to be no vaccine available for strongyloidiasis. Wearing shoes, particularly in known endemic regions, and keeping away from contact of unblemished skin with tainted discharges are the most normal approaches to avoid contamination with Strongyloides (Satoh & Kokaze, 2004; Ohnishi et al., 2004). Increased worm loads in patients with hyperinfection or on the other hand scattered malady could cause nosocomial transmission of the pathogen; along these lines, hospitalized patients ought to be isolated (Igra-Siegman et al., 1981; Satoh & Kokaze, 2004). Nosocomial transmission of contamination has not been accounted for, yet an instance of assumed individual to-individual transmission has been published (Czachor & Jonas, 2000). To forestall hyperinfection in patients as of now constantly tainted, the best technique might be to distinguish and treat these people before immunosuppressive treatment (eg, disease chemotherapy, immunosuppression for organ transplantation, unending steroids) is begun or hematopoietic undifferentiated organism transplantation is performed (Siddiqui & Berk, 2001; Safdar et al., 2004; Gupta et al., 2006; Marty et al., 2005). The life form could likewise be transmitted from the transplanted kidney of a tainted donor (Hoey et al., 1981). However, after the presentation of cyclosporine, which has been accounted for to have action against the parasite, there have been no instances of Strongyloides hyperinfection or spread infection in patients accepting cyclosporine following renal transplantation (Ghoshal et al., 2006; Liu & Weller, 1989). In the US, explicit treatment rules are accessible only for patients who have ventured out to endemic regions, have eosinophilia, and are to experience bone marrow transplantation, or on the other hand who have positive screening tests for Strongyloides. These patients ought to be treated with ivermectin agreeing to the rules of the Centers for Disease Control and Prevention (Centers for Disease Control and Prevention, 2004). Another choice to forestall hyperinfection or dispersed malady is to hypothetically treat foreigners with or without eosinophilia who move to non-endemic zones with a solitary portion of ivermectin or albendazole (Muennig et al., 2004; Nutman, 2006). For proposed treatment, ivermectin has been demonstrated to be cost-effective at the point when contrasted with albendazole, notwithstanding when controlled for various pervasiveness rates (Muennig et al., 2004).

WHY STRONGILOIDEASIS IS A NEGELCTED TROPICAL DISEASE

Neglected Tropical Diseases (NTDs) are a diverse group of tropical contaminations which are regular in low-pay populaces in develoing locales of Africa, Asia, and the Americas. They are brought about by an assortment of pathogens, for example, infections, microscopic organisms, protozoa and helminths. These sicknesses are appeared differently in relation to the enormous three irresistible illnesses (HIV/AIDS, tuberculosis, and Malaria), which for the most part get more noteworthy treatment and research funding (Hotez, 2013). In sub-Saharan Africa, the impact of these ailments as a group is practically identical to intestinal sickness and tuberculosis (Hotez & Kamath, 2009). NTD co-disease can likewise make HIV/AIDS and tuberculosis further lethal (Mike 2006). Inside developing nations, neglected tropical diseases influence the most deprived in the society. In the United States, there are up to 1.46 million families including 2.8 million youngsters living on under two dollars a day (Research Publications, 2018). In nations, for example, these, the hefts of neglected tropical diseases are frequently dominated by other general medical problems. Be that as it may, a considerable lot of similar issues put populaces in danger in developed as developing countries. For instance, different issues can come from destitution which unclutter people to the vectors of these ailments, for example, absence of sufficient housing (Hotez & Peter, 2012). Twenty neglected tropical diseases are ranked by the World Health Organization (WHO), nevertheless other organizations describe NTDs inversely.

The pervasiveness of *S. stercoralis* is believed to be harshly thought little of because of the low affectability of the current accessible symptomatic devices and the tools for the diagonosis. Morever the deficiency of specific studies also puts the disease in the neglected ones. The most generally utilized strategy for the recognizable proof of human intestinal helminth diseases in developing nations, for example the Kato-Katz system, identifies eggs in excrement in any case, neglects to show hatchlings. The exhibition of standard methodology like formol-ether concentration technique is restricted by the low convergence of larval stages in fecal specimens in incessant strongyloidiasis. Parasite-explicit fixation strategies like the Baermann system and culture methods (Cheesbrough 1987) are substantially more delicate, however even with the utilization of these techniques some light diseases are missed. Along these lines, *S. stercoralis* diseases are only here and there identified and numerous territories without data persevere on the worldwide *S. stercoralis* conveyance map. In this manner, the epidemiological learning with respect to this parasite is inconsistent, best case scenario. Just in patients with dispersed strongyloidiasis may huge quantities of hatchlings be exhibited in faeces, sputum, pee what's more, lung liquids just as in organ biopsies. Perpetual uncomplicated *S. stercoralis* diseases additionally cause unpretentious pathology, which is non-explicit for this contamination. In this manner, the contamination frequently stays undiscovered. Numerous epidemiological parts of Strongyloides contamination are obscure or just inadequately comprehended. We don't have the foggiest idea in detail where Strongyloides is endemic, which commonness rates and forces can ordinarily be normal in various settings and populaces, the zoonotic perspectives, land varieties, for example in transmission and clinical introduction, have invulnerability and the hazard factors for dispersal, and social, social, biological and natural angles. Little is thought about the genuine clinical importance of Strongyloides contaminations, the side effects related with distinctive disease designs, the relationship between Strongyloides and different irresistible illnesses, and of the connection between serious contaminations and immune system sicknesses for example, diabetes. The pertinence of ailing health for Strongyloides, particularly among youngsters, and the impacts of the subsequent state on the clinical result of *S. stercoralis* disease just as suggestions for pregnant ladies still should be uncovered.

CONCLUSION AND RECOMMENDATIONS

Strongyloides is a significant parasite that can cause serious malady in patients who become immunosuppressed quite a long while in the wake of securing the underlying contamination. Clinicians rehearsing in endemic zones or in territories with enormous vagrant populaces from endemic zones ought to be acquainted with this contamination, as patients may present with hyperinfection or disseminated sickness numerous years after the first introduction. The most normally utilized medicines incorporate ivermectin, thiabendazole, and albendazole. *Strongyloides stercoralis* has been the ignored and neglected member from the STH family for quite a long time, generally due to an absence of analytic precision and dilemmas in assessing public wellbeing mediations. The worldwide commonness and clinical effect of strongyloidiasis are probably going to be significantly thought little of.

There is a lot of that remained to be documented about *Strongyloides stercoralis*. The biology of the parasite, specifically its genomic possessions and the method of replication of the female parasitic worm, should be additionally clarified. We don't know in detail where Strongyloides is endemic, which commonness rates and intensities can regularly be normal in various settings and populaces, the zoonotic viewpoints, geological varieties, for example in transmission and clinical introduction, host immunity and the hazard factors for spread, and social, environmental and ecological angles. There likewise is a critical need to utilize current sub-atomic techniques to improve analysis and for the separation among species and strains from various hosts and topographical basis.

Little is understood about the genuine clinical pertinence of Strongyloides contaminations, the indications related with various disease designs, the relationship among Strongyloides and different irresistible ailments, and the connection between serious diseases and immune system disorders, for example, diabetes. The significance of malnutrition for Strongyloides, particularly among kids, and the impacts of the subsequent state on the clinical result of *S. stercoralis* contamination just as suggestions for pregnant ladies still should be uncovered.

Given the inclination to create extreme infection in certain populace subgroups, *S. stercoralis* is of chief public health concern. We have to prudently explore this parasite, create control projects and execute them where proper. Human strongyloidiasis is a cosmopolitan public health irresistible malady that is distinctive compared to other STH. These unique highlights must be addressed and assimilated in public health techniques for analysis, treatment and for individual case controlling.

REFERENCES

Abraham, D., Rotman, H. L., Haberstroh, H. F., Yutanawiboonchai, W., Brigandi, R. A., Leon, O., ... Schad, G. A. (1995). *Strongyloides stercoralis*: Protective immunity to third-stage larvae in BALB/cByJ mice. *Experimental Parasitology*, *80*(2), 297–307. doi:10.1006/expr.1995.1036 PMID:7895840

Adenusi, A. A. O., & Adenusi, A. O. (2003). Comparison of ivermectin and thiabendazole in the treatment of uncomplicated Strongyloides stercoralis infection. *African Journal of Biotechnology*, *2*(11), 465–469. doi:10.5897/AJB2003.000-1093

Alvarez, J. (1978). Strongyloidiasis en Guayaquil. Nuevas nociones epidemiologicas y biologicas. *Anales de Medicina y Cirugía*, *2*, 1–7.

Arroyo, R., Troper, L., & Vargas, G. (1985). Estrongiloidiasis fatal: Studio de ocho casos de autopsia. *Patología (México)*, *23*, 135–146.

Bannon, J. P., Fater, M., & Solit, R. (1995). Intestinal ileus secondary to Strongyloides stercoralis infection: Case report and review of the literature. *The American Surgeon*, *61*, 377–380. PMID:7893110

Barrett, K. E., Neva, F. A., Gam, A. A., Cicmanec, J., London, W. T., Phillips, J. M., & Metcalfe, D. D. (1988). The immune response to nematode parasites: modulation of mast cell numbers and function during Strongyloides stercoralis infections in nonhuman primates. *American Journal of Tropical Medicine and Hygiene* *38*: 574-581.

Berthoud, F., & Berthoud, S. (1975). A propos de 18 cas d'anguillulose diagnostiques a Geneve. *Schweizerische Medizinische Wochenschrift*, *105*, 1110–1115. PMID:1162316

Bisoffi, Z., Buonfrate, D., Sequi, M., Mejia, R., Cimino, R. O., Krolewiecki, A. J., ... Nutman, T. B. (2014). Diagnostic accuracy of five serologic tests for *Strongyloides stercoralis* infection. *PLoS Neglected Tropical Diseases*, *8*(1). doi:10.1371/journal.pntd.0002640 PMID:24427320

Boggild, A. K., Libman, M., Greenaway, C., & McCarthy, A. E. (2016). CATMAT statement on disseminated strongyloidiasis: Prevention, assessment and management guidelines. *Canada Communicable Disease Report*, *42*(1), 12–19. doi:10.14745/ccdr.v42i01a03 PMID:29769976

Breitenfeld, V., & Iskra, M. (1984). Strongiloidoza u nas i u svijetu. *Lijecnicki Vjesnik*, *106*, 33–38. PMID:6708716

Brindley, P. J., Gam, A. A., Pearce, E. J., Poindexter, R. W., & Neva, F. A. (1988). Antigens from the surface and excretions/secretions of the filariform larva of *Strongyloides stercoralis*. *Molecular and Biochemical Parasitology*, *28*(17), 1–180. PMID:3386679

Brumpt, Le., Sang, H. T., & Jaeger, G. (1972). Quelques reflexions apropos du parasitisme sanguin et intestinal dans deux villages d'Afrique Centrale. *Bulletin de la Société de Pathologie Exotique*, *65*, 263–270.

Buonfrate, D., Sequi, M., Mejia, R., Cimino, R. O., Krolewiecki, A. J., Albonico, M., ... Bisoffi, Z. (2015). Accuracy of five serologic tests for the follow up of Strongyloides stercoralis infection. *PLoS Neglected Tropical Diseases*, *9*(2). doi:10.1371/journal.pntd.0003491 PMID:25668740

Capell, F. S., Pujol, F. R., Garau, A. J., Pallares, G. R., & Campo, G. E. (1982). Hiperinfestacion por *Strongyloides stercoralis*. *Medicina Clínica*, *79*, 232–235. PMID:6983009

Carvalho, E. (1978). Strongyloidiasis. *Clinics in Gastroenterology*, *7*, 179–200. PMID:627098

Centers for Disease Control and Prevention. (2000). Infectious Diseases Society of America, American Society of Blood and Marrow Transplantation. Guidelines for preventing opportunistic infections among hematopoietic stem cell transplant recipients. *MMWR. Recommendations and Reports*, *49*(RR-10), 1–125.

Cheesbrough, M. (1987). *Medical laboratory manual for tropical countries* (2nd ed., Vol. 1). Cambridge: Tropical Health Technology and Butterworths.

Chen, J. J., Lee, C. M., & Changchan, C. S. (1994). Duodenal *Strongyloides stercoralis* infection. *Endoscopy*, *26*(02), 272. doi:10.1055-2007-1008965 PMID:8026386

Chiodini, P. L., Reid, A. J., Wiselka, M. J., Firmin, R., & Foweraker, J. (2000). Parenteral ivermectin in Strongyloides hyperinfection. *Lancet*, *355*(9197), 43–44. doi:10.1016/S0140-6736(99)02744-0 PMID:10615895

Conway, D. J., Bailey, J. W., Lindo, J. F., Robinson, R. D., Bundy, D. A. P., & Bianco, A. E. (1993). Serum IgG reactivity with 41-, 31-, and 28-kDa larval proteins of *Strongyloides stercoralis* in individuals with strongyloidiasis. *The Journal of Infectious Diseases*, *168*(3), 784–787. doi:10.1093/infdis/168.3.784 PMID:8354924

Conway, D. J., Lindo, J. F., Robinson, R. D., Bundy, D. A. P., & Bianco, A. E. (1994). *Strongyloides stercoralis*: Characterization of immunodiagnostic larval antigens. *Experimental Parasitology*, *79*(2), 99–105. doi:10.1006/expr.1994.1069 PMID:8056086

Crompton, W. T., & Tulley, J. J. (1987). How much ascariasis is there in Africa? *Parasitology Today (Personal Ed.)*, *3*(4), 123–127. doi:10.1016/0169-4758(87)90054-8 PMID:15462932

Czachor, J. S., & Jonas, A. P. (2000). Transmission of *Strongyloides steracolis* person to person. *Journal of Travel Medicine*, *7*(4), 211–212. doi:10.2310/7060.2000.00063 PMID:11003736

Datry, A., Hilmarsdottir, I., Mayorga-Sagastume, R., Lyagoubi, M., Gaxotte, P., Biligui, S., ... Gentilini, M. (1994). Treatment of *Strongyloides stercoralis* infection with ivermectin compared with albendazole: Results of an open study of 60 cases. *Transactions of the Royal Society of Tropical Medicine and Hygiene*, *88*(3), 344–345. doi:10.1016/0035-9203(94)90110-4 PMID:7974685

Delarocque-Astagneau, E., Hadengue, A., Degott, C., Vilgrain, V., Erlinger, S., & Benhamou, J. P. (1994). Biliary obstruction resulting from *Strongyloides stercoralis* infection. *Report of a case. Gut*, *35*(5), 705–706. doi:10.1136/gut.35.5.705 PMID:8200571

DeVault, G. A., King, J. W. Jr, Rohr, M. S., Landreneau, M. D., Brown, S. T. III, & McDonald, J. C. (1990). Opportunistic infections with *Strongyloides stercoralis* in renal transplantation. *Reviews of Infectious Diseases*, *12*(4), 653–671. doi:10.1093/clinids/12.4.653 PMID:2201067

Dreyer, G., Fernandes-Silva, E., Alves, S., Rocha, A., Albuquerque, R., & Addiss, D. (1996). Patterns of detection of *Strongyloides stercoralis* in stool specimens: Implications for diagnosis and clinical trials. *Journal of Clinical Microbiology*, *34*(10), 2569–2571. doi:10.1128/JCM.34.10.2569-2571.1996 PMID:8880521

Ejezie, G. C., & Otigbuo, I. N. (1987). The stabilising factors of parasitic infections in Nigeria. *Epidemiologie, Mikrobiologie, Imunologie*, *31*, 45–51. PMID:2883231

Faria, J. A. S. (1972). Prevalencia de helmintos em escolares de 7-15 anos na cidade de Salvador. *Revista da Sociedade Brasileira de Medicina Tropical*, *6*(5), 261–264. doi:10.1590/S0037-86821972000500004

Fiillebom, F. H. (1926). Autoinfektion bei Strongyloidestriigem. *Arch Schiffs Tropenhyg*, *30*, 721–732.

Forrer, A., Khieu, V., Schar, F., Hattendorf, J., Marti, H., Neumayr, A., ... Odermatt, P. (2017). *Strongyloides stercoralis* is associated with significant morbidity in rural Cambodia, including stunting in children. *PLoS Neglected Tropical Diseases*, *11*(10). doi:10.1371/journal.pntd.0005685 PMID:29059195

Gann, P. H., Neva, F. A., & Gam, A. A. (1994). A randomized trial of single- and two-dose ivermectin versus thiabendazole for treatment of strongyloidiasis. *The Journal of Infectious Diseases*, *169*(5), 1076–1079. doi:10.1093/infdis/169.5.1076 PMID:8169394

Genta, R. M. (1988). Predictive value of an enzyme-linked immunosorbent assay (ELISA) for the serodiagnosis of strongyloidiasis. *American Journal of Clinical Pathology*, *89*(3), 391–394. doi:10.1093/ajcp/89.3.391 PMID:3348175

Genta, R. M. (1989). Global prevalence of strongyloidiasis: Critical review with epidemiologic insights into the prevention of disseminated disease. *Reviews of Infectious Diseases, 11*(5), 755–767. doi:10.1093/clinids/11.5.755 PMID:2682948

Genta, R. M., Gatti, S., Linke, M. J., Cevini, C., & Scaglia, M. (1988). Endemic strongyloidiasis in northern Italy: Clinical and immunological aspects. *The Quarterly Journal of Medicine, 258*, 679–690. PMID:3256899

Genta, R. M., & Lillibridge, J. P. (1989). Prominence of IgG4 antibodies in the human responses to *Strongyloides stercoralis* infection. *The Journal of Infectious Diseases, 160*(4), 692–699. doi:10.1093/infdis/160.4.692 PMID:2794561

Ghoshal, U. C., Alexender, G., Ghoshal, U., Tripathi, S., & Krishnani, N. (2006). *Strongyloides stercoralis* infestation in a patient with severe ulcerative colitis. *Indian Journal of Medical Sciences, 60*(3), 106–110. doi:10.4103/0019-5359.22761 PMID:16569924

Giannoulis, E., Arvanitakis, C., Zaphiropoulos, A., Nakos, V., Karkavelas, G., & Haralambidis, S. (1986). Disseminated strongyloidiasis with uncommon manifestations in Greece. *The Journal of Tropical Medicine and Hygiene, 89*, 171–178. PMID:3783811

Grove, D. I. (1982). Treatment of strongyloidiasis with thiabendazole: An analysis of toxicity and effectiveness. *Transactions of the Royal Society of Tropical Medicine and Hygiene, 76*(1), 114–118. doi:10.1016/0035-9203(82)90034-7 PMID:7080143

Gupta, S., Jain, A., Fanning, T. V., Couriel, D. R., Jimenez, C. A., & Eapen, G. A. (2006). An unusual cause of alveolar hemorrhage post hematopoietic stem cell transplantation: A case report. *BMC Cancer, 6*(1), 87. doi:10.1186/1471-2407-6-87 PMID:16603072

Gustavo, S. J. (1993). Movement of the rhabditiform larva of *Strongyloides stercoralis. Lancet, 342*, 13–10. PMID:8100289

Holland, C. V., Crompton, D. W. T., Taren, D. L., Nesheim, M. C., Sanjur, D., Barbeau, I., & Tucker, K. (1987). Ascaris lumbricoides infection in pre-school children from Chiriqui Province, Panama. *Parasitology, 95*(3), 615–622. doi:10.1017/S0031182000058030 PMID:3696782

Hotez, P. (2012). Fighting neglected tropical diseases in the southern United States. *BMJ (Clinical Research Ed.), 345*(sep13 1). doi:10.1136/bmj.e6112 PMID:22977143

Hotez, P. J. (2013, November). NTDs V.2.0: "blue marble health"—neglected tropical disease control and elimination in a shifting health policy landscape. *PLoS Neglected Tropical Diseases*, *7*(11). doi:10.1371/journal.pntd.0002570 PMID:24278496

Hotez, P. J., & Kamath, A., & Cappello, M. (Eds.). (2009). Neglected Tropical Diseases in Sub-Saharan Africa: Review of Their Prevalence, Distribution, and Disease Burden. PLoS Negl Trop Dis. 3(8). doi:. doi:10.1371/journal.pntd.0000412

Igra-Siegman, Y., Kapila, R., Sen, P., Kaminski, Z. C., & Louria, D. B. (1981). Syndrome of hyperinfection with *Strongyloides stercoralis. Reviews of Infectious Diseases*, *3*(3), 397–407. doi:10.1093/clinids/3.3.397 PMID:7025145

Henriquez-Camacho, C., Gotuzzo, E., Echevarria, J., White Jr, A. C., Terashima, A., Samalvides, F., ... & Plana, M. N. (2016). Ivermectin versus albendazole or thiabendazole for *Strongyloides stercoralis* infection. *Cochrane Database of Systematic Reviews*, (1). PMID:26778150

Junod, C. (1987). Etude retrospective de 1,934 cas de strongyloidose diagnostiques aParis (1970-1986). Origine geographique, Epidemiologie. *Bulletin de la Société de Pathologie Exotique*, *80*, 357–369. PMID:3621398

Kennedy, S., Campbell, R. M., Lawrence, J. E., Nichol, G. M., & Rao, D. M. (1989). A case of severe *Strongyloides stercoralis* infection with jejunal perforation in an Australian ex-prisoner-of-war. *The Medical Journal of Australia*, *150*(2), 92–93. doi:10.5694/j.1326-5377.1989.tb136371.x PMID:2542737

Khieu, V., Schar, F., Marti, H., Sayasone, S., Duong, S., Muth, S., & Odermatt, P. (2013). Diagnosis, treatment and risk factors of *Strongyloides stercoralis* in schoolchildren in Cambodia. *PLoS Neglected Tropical Diseases*, *7*(2). doi:10.1371/journal.pntd.0002035 PMID:23409200

Krolewiecki, A. J., Lammie, P., Jacobson, J., Gabrielli, A.-F., Levecke, B., Socias, E., ... Albonico, M. (2013). A public health response against *Strongyloides stercoralis*: Time to look at soil-transmitted helminthiasis in full. *PLoS Neglected Tropical Diseases*, *7*(5), 21–65. doi:10.1371/journal.pntd.0002165 PMID:23675541

Lee, M. G., & Terry, S. I. (1989). Arteriomesenteric duodenal occlusion associated with strongyloidiasis. *The Journal of Tropical Medicine and Hygiene*, *92*, 41–45. PMID:2918578

Lima, J. P., & Delgado, P. G. (1961). Diagnosis of strongyloidiasis: importance of Baermann's method.*American Journal of Digestive Diseases 6 (New Series):* 899-904.

Lindo, J. F., Conway, D. J., Atkins, N. S., Bianco, A. E., Robinson, R. D., & Bundy, D. A. P. (1994). Prospective evaluation of enzyme-linked immunosorbent assay and immunoblot methods for the diagnosis of endemic *Strongyloides stercoralis* infection. *The American Journal of Tropical Medicine and Hygiene*, *51*(2), 175–179. doi:10.4269/ajtmh.1994.51.175 PMID:8074251

Little, M. D. (1966). Comparative morphology of six species of Strongyloides (Nematoda) and redefinition of the genus. *The Journal of Parasitology*, *52*(1), 69–84. doi:10.2307/3276396 PMID:5929983

Liu, L. X., & Weller, P. F. (1993). Strongyloidiasis and other intestinal nematode infections. *Infectious Disease Clinics of North America*, *7*, 655–682. PMID:8254165

Lo, C. Y., & Kajioka, R. (1989). Predictive values. *American Journal of Clinical Pathology*, *91*(4), 505–506. doi:10.1093/ajcp/91.4.505b PMID:2597242

Longworth, D. L., & Weller, P. F. (1986). Hyperinfection syndrome with strongyloidiasis. In J. S. Remington, & M. N. Swartz (Eds.), *Current clinical topics in infectious diseases* (Vol. 7, pp. 1–26). New York: McGraw-Hill.

Machicado, J. D., Marcos, L. A., Tello, R., Canales, M., Terashima, A., & Gotuzzo, E. (2012). Diagnosis of soil-transmitted helminthiasis in an Amazonic community of Peru using multiple diagnostic techniques. *Transactions of the Royal Society of Tropical Medicine and Hygiene*, *106*(6), 333–339. doi:10.1016/j.trstmh.2012.03.004 PMID:22515992

Marty, F. M., Lowry, C. M., Rodriguez, M., Milner, D. A., Pieciak, W. S., Sinha, A., ... Baden, L. R. (2005). Treatment of human disseminated strongyloidiasis with a parenteral veterinary formulation of ivermectin. *Clinical Infectious Diseases*, *41*(1). doi:10.1086/430827 PMID:15937753

Medina, L. S., Heiken, J. P., & Gold, R. P. (1992). Pipestem appearance of small bowel in strongyloidiasis is not pathognomonic of fibrosis and irreversibility. *American Journal of Roentgenology*, *159*(3), 543–544. doi:10.2214/ajr.159.3.1503020 PMID:1503020

Mejia, R., & Nutman, T. B. (2012). Screening, prevention, and treatment for hyperinfection syndrome and disseminated infections caused by *Strongyloides stercoralis*. *Current Opinion in Infectious Diseases*, *25*(4), 458–463. doi:10.1097/QCO.0b013e3283551dbd PMID:22691685

Messias, I. J. T., Genta, R. M., & Mohren, W. D. (1994). Adherence of monocytes and polymorphonuclear cells to infective larvae of *Strongyloides stercoralis* after complement activation. *The Journal of Parasitology, 80*(2), 267–274. doi:10.2307/3283757 PMID:8158470

Mike, S. (January 31, 2006). Beat neglected diseases to fight HIV, TB, and malaria. *SciDev.Net.* Archived from the original on May 19, 2006.

Molinari, G., Soloneski, S., & Larramendy, M. L. (2010). New ventures in the genotoxic and cytotoxic effects of macrocyclic lactones, abamectin and ivermectin. *Cytogenetic and Genome Research, 128*(1-3), 37–45. doi:10.1159/000293923 PMID:20389039

Muennig, P., Pallin, D., Challah, C., & Khan, K. (2004). The cost-effectiveness of ivermectin vs. albendazole in the presumptive treatment of strongyloidiasis in immigrants to the United States. *Epidemiology and Infection, 132*(6), 1055–1063. doi:10.1017/S0950268804003000 PMID:15635962

Nielsen, P. B., & Mojon, M. (1987). Improved diagnosis of *Strongyloides stercoralis* by seven consecutive stool specimens. Dntralblatt fur Bakteriologie. *Mikrobiologie und Hygiene, 263*, 616–618. PMID:3604502

Nutman, T. B. (2006). Asymptomatic peripheral blood eosinophilia redux: Common parasitic infections presenting frequently in refugees and immigrants [editorial]. *Clinical Infectious Diseases, 42*(3), 368–369. doi:10.1086/499247 PMID:16392082

Nutman, T. B., & Krolewiecki, A. (2019). Strongyloidiasis: A Neglected Tropical Disease. *Infectious Disease Clinics, 33*(1), 135–151. PMID:30712758

Ohnishi, K., Kogure, H., Kaneko, S., Kato, Y., & Akao, N. (2004). Strongyloidiasis in a patient with acquired immunodeficiency syndrome. *Journal of Infection and Chemotherapy, 10*(3), 178–180. doi:10.100710156-004-0312-8 PMID:15290458

Olsen, A., van-Lieshout, L., Marti, H., Polderman, T., Polman, K., Steinmann, P., ... Magnussen, P. (2009). Strongyloidiasis–the most neglected of the neglected tropical diseases? *Transactions of the Royal Society of Tropical Medicine and Hygiene, 103*(10), 967–972. doi:10.1016/j.trstmh.2009.02.013 PMID:19328508

Ordonez, L. E., & Angulo, E. S. (2004). Efficacy of ivermectin in the treatment of children parasitized by Strongyloides stercoralis. *Biomedica: revista del Instituto Nacional de Salud 24*(1): 33-41.

Ottesen, E. A., & Campbell, W. C. (1994). Ivermectin in human medicine. *The Journal of Antimicrobial Chemotherapy*, *34*(2), 195–203. doi:10.1093/jac/34.2.195 PMID:7814280

Politis, D. S., Triantafyllidi, E., Katsanos, K. H., & Christodoulou, D. K. (2017). Strongyloides stercoralis infection: An "exotic" cause of chronic diarrhea. *Annals of Gastroenterology*, *30*(4), 467. PMID:28655989

Requena-Mendez, A., Buonfrate, D., Bisoffi, Z., & Gutiérrez, J. M. (2014). Advances in the diagnosis of human strongyloidiasis. *Current Tropical Medicine Reports*, *1*(4), 440–475. doi:10.100740475-014-0034-7

Requena-Mendez, A., Chiodini, P., Bisoffi, Z., Buonfrate, D., Gotuzzo, E., & Muñoz, J. (2013). The laboratory diagnosis and follow up of strongyloidiasis: A systematic review. *PLoS Neglected Tropical Diseases*, *7*(1). doi:10.1371/journal.pntd.0002002 PMID:23350004

Research Publications. (July 23, 2017). Poverty Solutions at The University of Michigan. Retrieved from www.npc.umich.edu

Roger, D., & Gonzalez, E. M. N. (2017). Evaluacio'n del conocimiento sobre enfermdades de sexto curso del grado en Medicina de la Comunidad de Madrid. *IV Congreso de estudiantes de Medicina Universidad Rey Juan Carlos; May 25, 2017*. Madrid, Spain: Universidad Rey Juan Carlos.

Rotman, H. L., Yutanawiboonchai, W., Brigandi, R. A., Leon, O., Nolan, T. J., Schad, G. A., & Abraham, D. (1994). *Strongyloides stercoralis*: eosinophil dependent immune-mediated killing of infective L3 in Balb-kBYJ mice. *American Journal of Tropical Medicine and Hygiene. Abstracts:* 234.

Ruiz, M. J. A., Pons, M. J. A., Bermejo, L., Opez, J., Felices, A. F., Serra, S. J. A., & Felices, A. J. M. (1987). Estrongiloidiasis diseminada: Estudio clinico-patologico de un caso diagnosticado en autopsia. *Revista Clínica Española*, *180*, 381–384. PMID:3602552

Safdar, A., Malathum, K., Rodriguez, S. J., Husni, R., & Rolston, K. V. I. (2004). Strongyloidiasis in patients at a comprehensive cancer centre in the United States: A retrospective study covering the years 1971–2003. *Cancer*, *100*, 1531–1536. doi:10.1002/cncr.20120 PMID:15042689

Sandground, J. H. (1926). Biological studies on the life-cycle in the genus *Strongyloides grassi*, 1879. *American Journal of Hygiene*, *6*, 337–388.

Sandosham, A. A. (1952). An investigation into the association of creeping eruption with Strongyloides infection contracted in the Far East. *Journal of Helminthology*, *26*(1), 1–24. doi:10.1017/S0022149X00032569

Sato, Y., Inoue, F., Matsuyama, R., & Shiroma, Y. (1990). Immunoblot analysis of antibodies in human strongyloidiasis. *Transactions of the Royal Society of Tropical Medicine and Hygiene*, *84*(3), 403–406. doi:10.1016/0035-9203(90)90337-E PMID:2260175

Sato, Y., & Shiroma, Y. (1989). Peripheral lymphocyte subsets and their responsiveness in human strongyloidiasis. *Clinical Immunology and Immunopathology*, *53*(3), 430–438. doi:10.1016/0090-1229(89)90005-6 PMID:2572368

Satoh, M., & Kokaze, A. (2004). Treatment strategies in controlling strongyloidiasis. *Expert Opinion on Pharmacotherapy*, *5*(2), 293–301. doi:10.1517/14656566.5.11,2293 PMID:15500376

Scaglia, M., Brustia, R., Gatti, S., Bernuzzi, A. M., Strosselli, M., Malfitano, A., & Capelli, D. (1984). Autochthonous strongyloidiasis in Italy: An epidemiological and clinical review of 150 Cases. *Bulletin de la Société de Pathologie Exotique*, *77*, 328–332. PMID:6488423

Scowden, E. B., Schaffner, W., & Stone, W. J. (1978). Overwhelming strongyloidiasis. An unappreciated opportunistic infection. *Medicine; Analytical Reviews of General Medicine, Neurology, Psychiatry, Dermatology, and Pediatries*, *57*(6), 527–544. doi:10.1097/00005792-197811000-00004 PMID:362122

Siddiqui, A. A., & Berk, S. L. (2001). Diagnosis of *Strongyloides stercoralis* infection. *Clinical Infectious Diseases*, *33*(7), 1040–1047. doi:10.1086/322707 PMID:11528578

Spinicci, M., Macchioni, F., Mantella, A., Gabrielli, S., Roselli, M., Rojo-Mayaregua, D., ... Bartoloni, A. (2017). Seroepidemiological trend of strongyloidiasis in the Bolivian Chaco (1987–2013) in the absence of disease-specific control measures. *Tropical Medicine & International Health*, *22*(11), 1457–1462. doi:10.1111/tmi.12977 PMID:28929562

Sprott, V., Selby, C. D., Ispahani, P., & Toghill, P. J. (1987). Indigenous strongyloidiasis in Nottingham. *British Medical Journal (Clinical Research Ed.)*, *294*(6574), 741–742. doi:10.1136/bmj.294.6574.741-a PMID:3105720

Suputtamongkol, Y., Premasathian, N., Bhumimuang, K., Waywa, D., Nilganuwong, S., Karuphong, E., ... Silpasakorn, S. (2011). Efficacy and safety of single and double doses of ivermectin versus 7-day high dose albendazole for chronic strongyloidiasis. *PLoS Neglected Tropical Diseases*, *5*(5). doi:10.1371/journal.pntd.0001044 PMID:21572981

Thompson, J. R., & Berger, R. (1991). Fatal adult respiratory distress syndrome following successful treatment of pulmonary strongyloidiasis. *Chest*, *99*(3), 772–774. doi:10.1378/chest.99.3.772 PMID:1995245

Hoy, W. E., Roberts, N. J., Bryson, M. F., Bowles, C., Lee, J. C., Rivero, A. J., & Ritterson, A. L. (1981). Transmission of strongyloidiasis by kidney transplant? Disseminated strongyloidiasis in both recipients of kidney allografts from a single cadaver donor. *Journal of the American Medical Association*, *246*(17), 1937–1939. doi:10.1001/jama.1981.03320170049031 PMID:7026811

Vazquez-Guillamet, L. J., Saul, Z., Miljkovich, G., Vilchez, G. A., Mendonca, N., Gourineni, V., ... Gangcuangco, L. M. (2017). *Strongyloides stercoralis* infection among human immunodeficiency virus (HIV)-infected patients in the United States of America: A case report and review of literature. *The American Journal of Case Reports*, *18*, 339–346. doi:10.12659/AJCR.902626 PMID:28366929

Viera, R. A. (1979). *"Strongyloides stercoralis."* Um nematodo de relevante importancia em medicina humana. *Jornal do Medico C*, *1838*, 109–126.

WHO. (2017). *Preventive chemotherapy to control soil-transmitted helminth infections in at-risk population groups* (p. 87). Geneva, Switzerland: WHO.

Woodring, J. H., Halfhill, H. I. II, & Reed, J. C. (1994). Pulmonary strongyloidiasis: Clinical and imaging features. *AJR. American Journal of Roentgenology*, *162*(3), 537–542. doi:10.2214/ajr.162.3.8109492 PMID:8109492

Zaha, O., Hirata, T., Kinjo, F., & Saito, A. (2000). Strongyloidiasis--progress in diagnosis and treatment. *Internal Medicine (Tokyo, Japan)*, *39*(9), 695–700. doi:10.2169/internalmedicine.39.695 PMID:10969899

KEY TERMS AND DEFINITIONS

Ivermectin: Ivermectin is an antihelmintic drug, derived from avermectin and is used to treat enterobiasis and certain parasitic roundworm infections by paralysing or killing the parasites. It is an essential medicine recognised by the World Health

Organization as it has a favourable effect in improving the well-being of people throughout the world.

Neglected Tropical Disease: Neglected tropical diseases (NTDs) are a diverse pool of tropical ailments which generally distress the world's poor people and truly have not gotten as a lot of consideration as other ailments. They are brought about by a variety of pathogens, for example viruses, bacteria, microbes, protozoa and helminths.

Rhabditiform: It is the initial developmental larval stage (first and second) of soil-borne nematodes for example *Ancylostoma*, *Necator* and *Strongyloides*, which precede the infectious third-stage filariform larva. Rhabditiform means "not comparable" and having the character of a nematode of the order Rhabditida.

Stercoralis: *Stercoralis* is a species of the *Strongyloides* genus. It is a human pathogenic parasitic roundworm causing strongyloidiasis. Its common name is threadworm or pinworm. The adult parasitic stage lives in canals of the mucosa of small intestine.

STH: STH is read as "Soil-Transmitted Helminths" as well as "Soil Transmitted Helminthiasis". These are the intestinal worms which are transmitted to humans through contaminated soil. The main species that infect humans are *Ascaris lumbricoides* (roundworm), *Trichuris trichiura* (whipworm), *Necator americanus* (new world hookworm) and *Ancylostoma duodenale* (hookworm).

Stool: Stool, the other name for human feces or excrement, is the solid matter which is discharged from large intestine through the anus during defecation. To diagnose a medical condition, a stool sample may be sometimes required for the testing purposes.

Strongyloidiasis: Strongyloidiasis is a human parasitic ailment brought about by a whip like nematode worm called *Strongyloides stercoralis*. It is a condition which affects numerous organ frameworks, including the central nervous system that can prompt death if untreated.

Thiabendazole: Thiabendazole is a benzimidazole derivative with anthelminthic property. Thiabendazole is a broad spectrum antihelmintic agent used primarily in treatment of intestinal pinworm and strongyloides infection.

Chapter 6
The Role of Community Participation in Fighting Schistosomiasis:
Lessons From Akka Oases (Southern Morocco)

Adil Moumane
https://orcid.org/0000-0003-0296-2679
Ibn Tofail University, Kenitra, Morocco

Ahmed Karmaoui
https://orcid.org/0000-0003-3881-4029
Southern Center for Culture and Sciences, Zagora, Morocco

Jamal Al Karkouri
Ibn Tofail University, Kenitra, Morocco

Jamal Akchbab
Friends of the Environment Organization, Zagora, Morocco

ABSTRACT

In 2009, the Moroccan Ministry of Health in collaboration with the World Health Organization confirmed that the transmission of schistosomiasis had ceased in all previously endemic provinces of Morocco. This achievement couldn't have been achieved without the effective participation and engagement of local communities. In this chapter, studies of community participation strategies for schistosomiasis control in Akka oasis were reviewed. Southern Morocco was known to be a foci site for urinary schistosomiasis since 1937. Combined efforts between health officials and local community participation have achieved the elimination of this neglected tropical disease, where the last cases in these foci were recorded in the early 2000s. The results of this chapter can be used to increase awareness of the need for community participation in controlling neglected tropical diseases such as schistosomiasis.

DOI: 10.4018/978-1-7998-2197-7.ch006

INTRODUCTION

Human communities have an extraordinary capacity to adapt and be resilient to their environments, for example by limiting their exposure to hazards and parasites and to take collective measures to protect themselves against hostile conditions (Schaefer 1993:1). Community involvement is one of the principles of the primary health care framework formulated at the Alma Ata conference in 1978 (WHO and UNICEF 1978). Besides, community participation is a fundamental principle of *Agenda 21*, the action document of the 1992 United Nations Conference on Environment and Development (UNCED), which specifies "mobilizing communities for action as essential in solving health, environment and development problems" (Schaefer 1993:19). Community involvement is well-recognized as an important tool for creating a healthier and more sustainable planet by both environment and health fields (Chu 1997). Moreover, several disease elimination campaigns achieved success due to the significant role of community involvement (Atkinson et al 2011) such as the elimination of schistosomiasis in Morocco.

Foci of schistosomiasis were widespread in Morocco. Urinary schistosomiasis was the form of schistosomiasis that spread in the country, which was caused by the *Schistosoma haematobium* blood parasite (Doumenge, et al. 1987: 22). Schistosomiasis had been endemic in the oasis zone of southern Morocco for a lot of years. Oasis regions of Morocco, by their geographical position, climatic conditions, demographic, and socio-economic system, have a lot of the conditions associated with the development of endemic neglected tropical diseases (NTD) such as urinary schistosomiasis and cutaneous leishmaniasis (Karmaoui 2018).

Since 1982, the schistosomiasis control program was integrated into the Moroccan health system and started operating in 1983 in all 20 provinces where schistosomiasis foci were reported (Laaziri and Benouna, 1982). A program to eradicate schistosomiasis over all the endemic geographic regions started in 1994. The disease was eradicated at the national level after ten years: no new cases have been reported since 2004. In 2009, schistosomiasis disease was considered to be eliminated by the World Health Organization and the Moroccan Ministry of Health in all previously endemic provinces (Laaziri 2012). This achievement wouldn't have been possible without effective participation and engagement of local communities.

Akka oases in Southern Morocco were known to be an ancient foci site for urinary schistosomiasis since 1937 (Direction de la santé public 1940:20). After a long struggle for the local community, combined efforts between health officials and local community have achieved the elimination of this NTD, where the last cases in these foci were recorded in the early 2000s. The aim of this study is to review community participation strategies for schistosomiasis used to control the disease in Akka oasis. The resulting information can be used to increase awareness of the need for community participation in controlling neglected tropical diseases.

Study Area

Akka Oases, Tata province, (Figure 1, 3) are situated on the alluvial plain of *Oued* Akka, which is one of the tributaries of the Draa valley basin. Located in the pre-Saharan zone of southern Morocco, it is surrounded by the Jbel Bani mountains and limited on the south by the Oued Draa. *The* Akka river crosses the Jbel Bani mountains through narrow gorges called *foum*, which collect water and have made irrigated agriculture possible in this oasis zone of the southern slopes of *Jbel* Bani (Oudada 2009). The climate is arid, with an average annual rainfall of less than 60mm per year (DPE), and maximum temperatures that can exceed 50 ° C (Steer L et al 2004). Agriculture depends on traditional irrigation to cover an area of approximately 1530 ha. The population in the palm oasis was estimated at 6 132 habitants in 2004 (HCP 2004). Small cross-sectional dams have been constructed

Figure 1. A. Geographic location of Draa valley basin in Morocco, B. Historical urinary schistosomiasis foci in the Draa valley basin within Akka Oases (Morocco).
Data Source: WHO 1987 in Doumenge et al. 1987: 27, modified

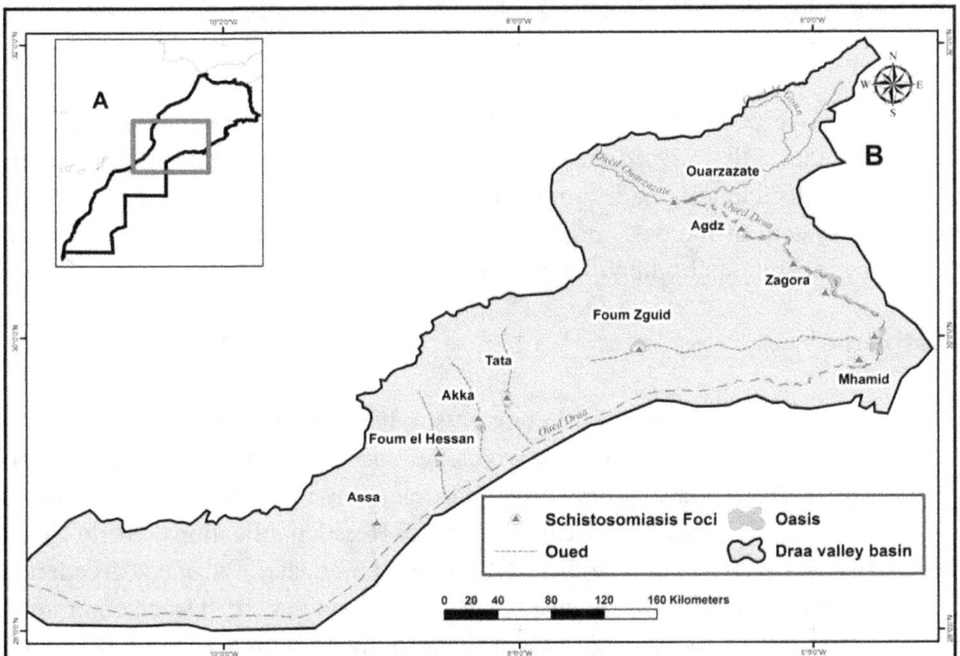

Figure 2. Akka oasis's Irrigation systems compounds, A: Akka's dam. B:Seguia: Irrigation canal . C:Amond Oasis pond.
Source: Authors

in *the* Akka riverbed to impound water. Water from ponds is directed into the major channel system called *Seguia* and conveyed through a network of small canals to the fields by small channels, made of mud, to supply water to the fields. The traditional irrigation system of the oasis regions is based on the construction and the use of *Seguia* by inhabitants of the villages (Figure 2).

Methods

The present paper is a result of several visits to the ancient schistosomiasis foci of Akka Oases, and an extensive review which has documented the impact and lessons learned from Akka Oases community participation in schistosomiasis control. Databases reviewed were: Moroccan Ministry of Health publications, World Health Organization (WHO) publications, NCBI, Researchgate, PubMed, Academia, and Google Scholar. Keywords used in the search included; 'Oases community participation OR community Akka', 'Schistosomiasis Morocco or Health report

Morocco', and 'Elimination Schistosomiasis.' Both English and French language publications were used.

Water Use and Schistosomiasis Transmission

In Morocco, schistosomiasis (Bilharziasis) is caused by the infectious trematode larvae *Schistosoma haematobium*, where the *Bulinus truncatus* species of freshwater snails is the obligatory intermediate hosts of this parasite (Doumenge et al. 1987: 21-7 and Yacoubi et al.2007). People are infected through repeated contact with

Figure 3. Distribution of schistosomiasis cases by villages (Foci) within the oases of Akka (Morocco) the numbers between brackets indicate the cases recorded in each Oasis during the period between 1997 and 2003
Data Source: Laaziri 2012: 135 and Map: Authors

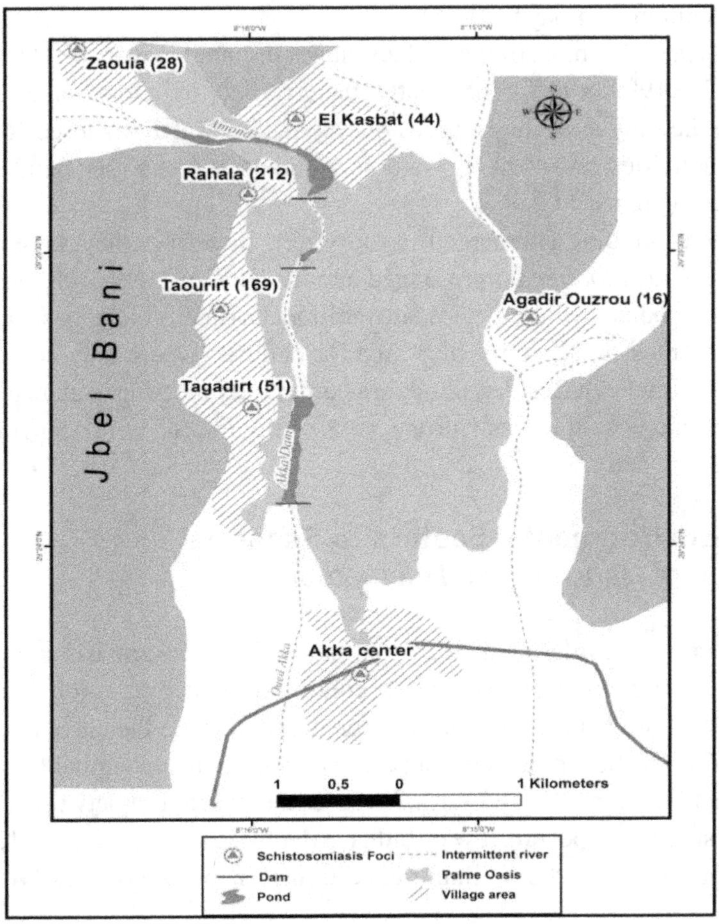

freshwater contaminated with the parasite and the infection rate was higher in bodies of water where the intermediate host breeds (Rozendaal 1997).

The distribution of cases in the study area shows that nearly three-quarters of cases are recorded in two villages: Rahala (40.8%) and Taourirte (32.5%). Both are located near permanent ponds, called Amond and Akka's dam respectively (Figure 3).

Figure 4 presents the evolution of schistosomiasis cases detected in the province of Tata, during the period 1994–2008. Akka oases have the highest number of cases in Tata province, more than four-fifths of the cases between 1997 and 2003 and almost (98.6%) in 2003 (Laaziri 2012: 124,135). This figure shows fluctuations in the number of cases between 1994 and 2004. In 1994, the number of cases was 117 after an important decrease in 1995 with just 43 cases, the number of infected cases increased again; until it reached its highest value of 233 cases in 1997. After 1997, a steady decline of new schistosomiasis cases over the years was recorded.

The community probably had been infected with schistosomiasis because of the use of Akka river ponds; where large populations of the intermediate snail host prosper (Laamrani & Lhayati 1996).

Until the late 90s, most of the Akka villages did not have drinkable tap water at home (Boelee 1999: 49). Consequently, the riverbed, irrigation canals, and a few wells were the only available sources of water. This water was used for drinking, household activities, and drinking water for animals. Women used ponds and canals to do laundry (Figure 5 C).

They were standing submerged in the water; therefore, they were exposed to *Schistosoma haematobium* for prolonged periods (Boelee 1999 p 48). Farmers were also always in contact with water used in traditional irrigation schemes. Furthermore, ponds represented an attractive play area for children where they could play, fish and swim (Figure 5 A). Besides, ponds are community meeting centers for villagers, and small markets selling oasis products were organized daily near *Amond* pond (Figure 5 B).

Linking Environmental Factors to Seasonal Variations of *Schistosoma Haematobium*

The presence of the intermediate carrier snail of schistosomiasis in a freshwater ecosystem depends on several ecological conditions; Breeding colonies are found in permanent and stagnant waters, or in water streams where the current is very slow (WHO 1957). The host prefers water polluted by human and animal faeces or urine near villages (Watson 1958: 839-40). Water at a temperature between 18 to 26 °C, with direct sunlight exposure, low turbidity, a depth that doesn't exceed 1.5-2 meters and a pH of 6 or 7 are the optimum conditions for these molluscs (WHO 1957).

Figure 4. Evolution of schistosomiasis Cases detected in the province of Tata during the period 1994–2008.
Data source: Laaziri 2012 p 124,135

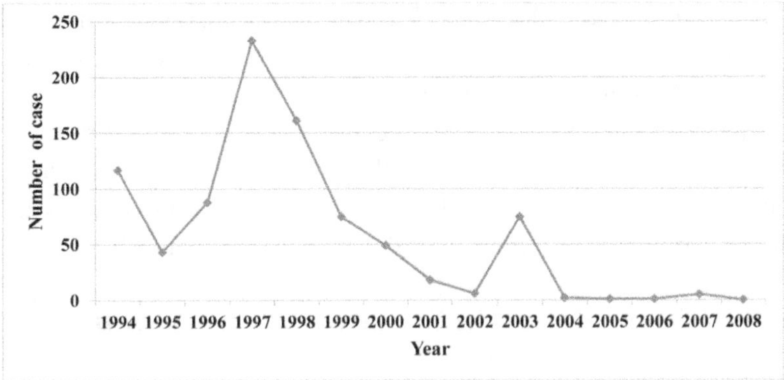

Figure 5. Different water uses and contacts by Akka'community. A: Kids swimming at Amound pond B: Amond pond is a meeting place and center of social life C: Women in a laundry site in Oued Akka riverbed
Source: Authors

Variations in these conditions explain the seasonal and geographical variations in host snail abundance in time and space (Doumenge 1987).

B. truncatus was active all year in Morocco, but during the summer snail populations density were increased (Khallaayoune & Laamrani 1992), which coincided with the highest period of schistosomiasis infection. Transmission varied from year to year related to the rainfall distribution. (idem)

Floods are destructive to this snail because of the sudden drop in temperature which often accompanies sudden floods (Watson 1958). Moroccan Saharan regions are characterized by autumn rain that causes violent flooding of rivers and streams leading to the temporary elimination of the snails (Doumenge 1987).

Community Participation in Schistosomiasis Control

B. truncatus snails were common and widespread throughout water collected in Akka, especially in ponds and canals. The limiting ecological factor was the density of aquatic vegetation, which can reduce water velocities, provide shelter from excessive direct sunlight and also offer support for egg masses (Boelee and Laamrani 2004).

According to the results of a participatory rapid appraisal about schistosomiasis organized in May 1997 in Akka oasis, participants from the local community proposed several interventions that can be used for schistosomiasis control.

These included measures to reduce contact with water, mechanical weed removal in transmission sites, and campaigns in schools since children often pass on the message to the household better than adults... (See Table 1 and Boelee 1999: 50). The community identified snail control as a strategy to reduce the transmission of schistosomiasis (Boelee and Laamrani 2004). Understanding the ecology of snails in local settings was the entry point for community involvement (Bruun & Aagaard-Hansen 2008). The density of aquatic vegetation in the water, especially *Potamogeton sp.,* was identified as a key factor determining snail existence and abundance (Laaziri 2012).

Communities repeatedly cleaned *seguias* several times between 1997 and 1998, which involved removal of vegetation and sediments, creating a smoother canal bed and steeper banks, allowing for higher water flow velocities. These efforts led to a significant reduction in density of snails and their egg masses. The authors Boelee & Laamrani (2004) concluded that environmental control measures were highly effective in reducing populations of *B. truncates,* and that the active role of the local community that designed, carried out and funded the measures without any outside support was essential in this achievement.

*Table 1. Recommendations for schistosomiasis control, as proposed by inhabitants of the villages of Irhalen, Taourirt, and Kasbah in the Akka oasis. The suggestions marked with * were put into practice less than a year after being formulated.*

Type of Intervention	Actions
Reduction of water contact	• Fence off the river bed • Warning signs at infested sites (design provided through school contest) • Appoint a guard • Penalty system for children playing in the water • Installation of convenient obligatory passages over the river • Symbolic fine for crossing elsewhere after the construction of new passages • Installation of taps • Construction of safe laundry sites • Installation of safe recreation basins in the villages
Snail control	• Reconstruction of the springs • Chemical snail control (despite the noxious effect on frogs and fish) • Mechanical weed removal in transmission sites • Reinforce collaboration between village committees and spring committees
Health education	• Make a video documentary in the local language with local images • Plan education campaigns during the summer, when water contact and contamination risk are highest • Campaigns in schools, as children can pass on the message to the household better than adults • Make village leaders aware of their responsibility to inform parents • Target information at women since they are more exposed to schistosomiasis and are less exposed to regular campaigns • Explain that after several negative screenings, it is still worthwhile and important to participate in case detection campaigns the following year • Involve sectors other than health and interior in education
Other Ministry of Health activities	• Check urine during each visit to the health post • Add a nurse to the schistosomiasis elimination team activities • Have medical doctor based in the Kasbah health post • Integrate schistosomiasis control activities with the trachoma control program

Source: (Boelee E 1999: 50).

DISCUSSION AND CONCLUSION

Between 1994 and 1997, most cases of schistosomiasis were recorded in two villages called Rahala and Tagadirt (Figure 3). This result may be explained by their proximity to big water ponds in the studied area. Fewer cases were also recorded in Agadir Ouzrou village, which is the furthest village from water ponds in this region. During our visits to Akka oases, we were told by local sources that the village uses underground water for irrigation, via a system of underground canals called khetara. Therefore, this may confirm what authors Laamrani & Lhayati (1996) concluded, that schistosomiasis infection in those oases was due to the use of water where large populations of the host snail *Bulinus truncatus* prosper.

Although the local community was informed through official health education programs about risks related to using water with favorable ecological conditions for the development of the host snail, they continued to use it. This was because they could not find another water source in the arid pre-Sahara region. Prevalence and intensity of schistosomiasis were higher among children under 14 years old, representing 80,5% of total cases of infection, and most cases occurred during the summer (Laaziri 2012:135, Laamrani et al. 2000a). We assume that during the summer school holiday; ponds and streamlets in the riverbed were the only fresh and attractive area during the hot season in this area, and also there were no other alternative places for kids to play. Thus, an increase in cases of infection was recorded during this time, when the temperature for fecundity and growth of the intermediate host is optimal. A rise in temperature allows population size to increase as is confirmed by authors Kalinda et al,(2017).

After the Bilharzias Elimination Program (PEB) was launched in 1994 (Laamrani et al.2000 b), it operated with a vertical approach in Akka Oases. Chemotherapy treatment and chemicals were used for intermediate host control. A mass treatment campaign was planned without prior diagnosis, where the entire population of the villages surrounding Akka received medication against schistosomiasis in 1995 (Boelee 1999:47). This may explain the drop in numbers of cases between 1994 and 1995 observed in the graph (Figure 3), that decline in the apparition of new cases was due to the success of the PEB therapeutic approach that temporarily interrupted transmission. However, less than one year later, there was an increase in new cases so the success of this vertical approach was limited and unsustainable (Figure 4 and Boelee 1999: 47).

Since 1997, the use of a horizontal or 'bottom-up' approach (Atkinson-et al.2011) aims to engage and support the participation of the Akka community in identifying and prioritizing their concerns when making decisions about schistosomiasis control. Professionals and local authorities are then asked to support. this 'bottom-up' approach to participation which is valuable for creating low-cost, eco-friendly control and a sustainable decline of schistosomiasis cases until elimination. Costs were low and benefits were not limited to snail and schistosomiasis control since irrigation performance was also improved and environmental contamination by the application of pesticides in this vulnerable ecosystem was reduced (Boelee and Laamrani 2004).

As a conclusion, in Akka oasis the synergy between official health programs and community participation was successful. There are lessons to be learned from the elimination of schistosomiasis from Akka oasis: early involvement of local community participation in problem definition and identification stimulated the development of a sustainable and locally-appropriate intervention for the control of the snail hosts of Bilharziasis; by a permanent alteration of their habitats and control of their environment conditions. These interventions require an understanding of intermediate snail ecology and the use of mass community participation and methods adapted to local conditions. An integrated approach including the application of several measures and interventions proved to be effective in Akka oasis, and this approach is needed to control the transmission of schistosomiasis in other regions of the world.

ACKNOWLEDGMENT

We thank Sandrine Ceurstemont for reading this chapter and for her valuable comments.

REFERENCES

Atkinson, J. A., Vallely, A., Fitzgerald, L., Whittaker, M., & Tanner, M. (2011). The architecture and effect of participation: A systematic review of community participation for communicable disease control and elimination. Implications for malaria elimination. *Malaria Journal*, *10*(1), 225. doi:10.1186/1475-2875-10-225 PMID:21816085

Boelee, E. (1999). Irrigation ecology of schistosomiasis: environmental control options in Morocco. (PhD thesis), Wageningen University, The Netherlands.

Boelee, E., & Laamrani, H. (2004). Environmental control of schistosomiasis through community participation in a Moroccan oasis. *Tropical Medicine & International Health*, *9*(9), 997–1004. doi:10.1111/j.1365-3156.2004.01301.x PMID:15361113

Bruun, B., & Aagaard-Hansen, J. (2008). *The social context of schistosomiasis and its control: an introduction and annotated bibliography*. World Health Organization.

Chu, C. (1994). *Community participation in public health: Definitions and conceptual framework. s, Ecological Public Health: From Vision to Practice*. Brisbane: Institute of Applied Environmental Research.

D. P. E. La Direction Provinciale de l'Équipement (*DPE*) de TATA (Poste D'AKKA) Service Eau.

Direction de la santé publique. (1940). BULLETIN DE L'INSTITUT D'HYGIÈNE DU MAROC.

Doumenge, J. P., Mott, K. E., Cheung, C., Villenave, D., Chapuis, O., Perrin, M. F., & Reaud-Thomas, G. (1987). Atlas of the global distribution of schistosomiasis. World Health Organization, Parasitic Diseases Programme. Presses Universitaires, Bordeaux, France.

HCP. (2004). HautCommissariat au Plan du Maroc, Recensement général de la population et de l'habitat 2004.

Kalinda, C., Chimbari, M. J., & Mukaratirwa, S. (2017). Effect of temperature on the Bulinusglobosus—Schistosomahaematobium system. *Infectious Diseases of Poverty*, *6*(1), 57. doi:10.118640249-017-0260-z PMID:28457230

Karmaoui, A. (2018). The cutaneous leishmaniasis vulnerability index (CLVI). *Acta Ecologica Sinica*, *38*(4), 288–295. doi:10.1016/j.chnaes.2018.01.001

Khallaayoune, K., & Laamrani, H. (1992). Seasonal patterns in the transmission of Schistosoma haematobium in Attaouia, Morocco. *Journal of Helminthology*, *66*(2), 89–95. doi:10.1017/S0022149X00012645 PMID:1640092

Laamrani, H., Khallaayoune, K., Madsen, H., Mahjour, J., & Gryseels, B. (2000a). New challenges in schistosomiasis control in Morocco. *Acta Tropica*, *77*(1), 61–67. doi:10.1016/S0001-706X(00)00114-5 PMID:10996121

Laamrani, H., & Lhayati, M. (1996). *Analyse préliminaire de la situation de la schistosomiase au foyerde Tata-Akka*. Rabat: Ministere de Sante Publique.

Laamrani, H., Mahjour, J., Madsen, H., Khallaayoune, K., & Gryseels, B. (2000b). Schistosoma haematobium in Morocco: Moving from control to elimination. *Parasitology Today (Personal Ed.)*, *16*(6), 257–260. doi:10.1016/S0169-4758(00)01665-3 PMID:10827435

Laaziri, M.(2012).Élimination de la schistosomiase au Maroc: Une réalité et un succès après trois décennies de lutte, OMS Bureau régional de la Méditerranée Orientale.

Laaziri, M., & Benouna, M. (1982). *Guide de la lutte contre la bilharziose. Direction des affaires techniques*. Rabat, Maroc: Ministe`re de la Sante Publique.

Oudada, M. (2009). L'eau dans le Bani: Disparités et gestion participative. In *Eaux, pauvreté et crises sociales*. Marseille: IRD Éditions; doi:10.4000/books.irdeditions.4893

Rozendaal, J. A. (1997). *Vector control: methods for use by individuals and communities*. World Health Organization.

Schaefer, M. (1993). *Health, environment, and development. Approaches to drafting country-level strategies for human well-being under Agenda 21 (No. WHO/EHE--93.1).* World Health Organization.

Steer, L., Goudet, M., Akchour, M., & Mouradi Abdellah, H. (2004). *Les plantes aromatiques, médicinales et tinctoriales, un atout pour le développement rural de la région de Tata. Centre National d'Etudes Agronomiques des Régions Chaudes. En collaboration avec Direction Provinciale de l'Agriculture de Tata et ALCESDAM (Maroc). Edition: P. JOUVE, C.* France: SEUGE.

Watson, J. M. (1958). Ecology and distribution of Bulinus truncatus in the Middle East; with comments on the effect of some human activities in their relationship to the snail host on the incidence of bilharziasis haematobia in the Middle East and Africa. *Bulletin of the World Health Organization, 18*(5-6), 833–894. PMID:13573115

WHO. (1957). Study Group on the Ecology of Intermediate Snail Hosts of Bilharziasis, World Health Organization, Study Group on the Ecology of Intermediate Snail Hosts of Bilharziasis [meeting held in Paris from October 3 to 9, 1956] : report. World Health Organization technical report series; No. 120.

WHO and UNICEF. (1978). Declaration of Alma-Ata. *WHO Chronicle, 32*, 428–430. PMID:11643481

Yacoubi, B., Zekhnini, A., Rondelaud, D., Vignoles, P., Dreyfuss, G., Cabaret, J., & Moukrim, A. (2007). Habitats of Bulinus truncatus and Planorbarius metidjensis, the intermediate hosts of urinary schistosomosis, under a semiarid or an arid climate. *Parasitology Research, 101*(2), 311–316. doi:10.100700436-007-0500-4 PMID:17340142

Chapter 7
Geographical Distribution of Cutaneous Leishmaniasis and Its Relationship With Climate Change in Southeastern Morocco

Abdelkrim Ben Salem
https://orcid.org/0000-0002-2283-5928
Department of Biology, Faculty of Sciences, University Mohammed V, Rabat, Morocco

Ahmed Karmaoui
https://orcid.org/0000-0003-3881-4029
Southern Center for Culture and Sciences, Zagora, Morocco

Souad Ben Salem
LHEACG (Faculty of Science Semlalia), Cadi Ayyad University, Marrakech, Morocco

Ali Ait Boughrous
Biology, health, and Environment Research Team, Facultés des Sciences et Techniques d'Errachidia, Morocco

ABSTRACT

The current chapter deals with one of the most neglected tropical diseases in Morocco, the cutaneous leishmaniasis. It is based on 10-year research (2010-2017) on the evolution of leishmaniasis taking climate change into account. Epidemiological and climatological data were collected from different administrations. The Geographic Information System (GIS) is chosen for interpolation, space-time analysis of climate data and map creation. The SPSS software was used for statistical analysis and to establish the relationship between Leishmaniasis and climatic conditions. Results show that the maximum number of cases is recorded in 2010 with 4,407 people affected while the low number is recorded in 2014 with 18 cases. Results also show a clear link between climatic factors and the incidence of the disease. The distribution of the disease in the province is influenced by maximum temperature, aridity, and vegetation cover. Additionally, anthropogenic factors play a significant role in explaining the emergence or re-emergence of leishmaniasis in the region.

DOI: 10.4018/978-1-7998-2197-7.ch007

INTRODUCTION

Leishmaniasis is a widespread parasitic disease that is transmitted to humans by the female bite of a hematophageal dipteral insect called phlebotoma or sandfly. The parasite is a flogged protozoan of the genus *Leishmania* that infects the mononucleated phagocytic system. Transmission occurs when the phlebotomus bites a sick man or animal and absorbs blood monocytes or parasitized dermal histiocytes. This vector stings another healthy man or animal and they will also be infected with the parasite, the latter will have as a reservoir of nomadic mammals including humans (Rguioui, 2006).

Leishmaniasis was ranked by the World Health Organization (WHO) among the main diseases to public health problems worldwide (Sadqi, 2013). In fact, leishmaniasis is endemic in 88 countries throughout Latin America, Africa, Asia and Southern Europe. 350 million people are at risk of infestation with a prevalence of 12 million people per year (Boussaa, 2008).

In Morocco, according to Ministerial Order 683-95- of March 31, 1995, leishmaniasis is a notifiable disease (MS, 2019). In this country, cutaneous leishmaniasis (CL) is endemic (Mahjour, Akalay, & Saddiki, 1992) and is widely distributed in three nosogeographic entities, zoonotic cutaneous leishmaniasis (ZCL; caused by *L. major*) located in the arid regions along the northern edge of the Sahara desert, anthroponotic cutaneous leishmaniasis (ACL; caused by L. *tropica*) in the semi-arid regions of central and south-western Morocco, and CL caused by *L. infantum* in the northern regions of the country (Rguioui, 2006). In the last terthy years, an increasing trend of geographic expansion was observed and new foci in several provinces of Morocco were recorded. A total of 24 804 cases of *L. major* and 16 852 cases of *L. tropica* were registered between 2004 and 2013 (Kahime et al., 2016).

The current study was carried out in the province of Errachidia that has a Mediterranean climate with dry summer according to the classification of Koppen-Geiger (Hufty, 2001). According to Regional Directorate of Health Draa-Tafilalet (SIAP), this area is affected by major cutaneous leishmaniasis (MCL) and more than 1900 cases have been diagnosed in 2017. These patients received free treatment in the various medical centers and facilities.

In this chapter, a retrospective analysis of the CL in Morocco was conducted based on epidemiological data of the period 2010-2017. The main objective was to highlight the relationship between climatic and environmental conditions and the distribution of CL by *L. major*.

MATERIAL AND METHODS

Study Area

The province of Errachidia (Figurc 1) covers an area of 42852 km^2 and has a population of nearly 418451 people according to the national census of 2014 (RGPH 2015). Geographically, it is demarcated by the province of Midelt, to the north-east by the province of Figuig, to the south and southeast by Algeria and to the west by the two provinces of Tinghir and Zagora.

After the creation of the two new provinces, Midelt and Tinghir, the number of municipalities in the province of Errachidia became 29 municipalities (7 urban centers and 22 rural centers).

The climate of Errachidia province is characterized by apparent variability in temperature (Figure 2) and humidity (semi-desert climate), as well as relatively low rainfall throughout the year. Very high summer temperatures (31.5°C as average to the month of July) and very low temperatures of winter (5°C as average in January). Low rainfall and irregularly distributed over time and space. Most of the territory is less than 100 mm of rain per year (Figure 3). However, the wind speed is above 57.6 km/h in May, June, July, and in August.

Data Collection

Epidemiological data were obtained from the registers and reports published by the local and national medical services. Monthly records of the number of LCM cases in the province of Errachidia between January 2010 and December 2017 were collected from the health service of the provincial health delegation. Monthly weather data between 1975 and 2016 were provided by the Rhris, Guir, Ziz, and Maider watershed of the Hydraulic Agency (ABHRGZ). This climatic data include minimum and maximum temperature, precipitation, relative humidity, evaporation, and wind speed of the 24 hydro-meteorological stations.

Tools and Approaches

Geographic Information System and Statistical Analysis

Arc-Gis 10.4 software was used for the map creation and analysis of climatic data. However, the statistical method used to establish the Leishmaniasis-climate relationship is Pearson's bilateral correlation with two levels of significance (0.01 and 0.05). IBM SPSS Statistics 19 is the software used for this statistical analysis. The correlations between the number of CML cases and each of the climate, aridity

Figure 1. Localization of the study area
Source: (Authors)

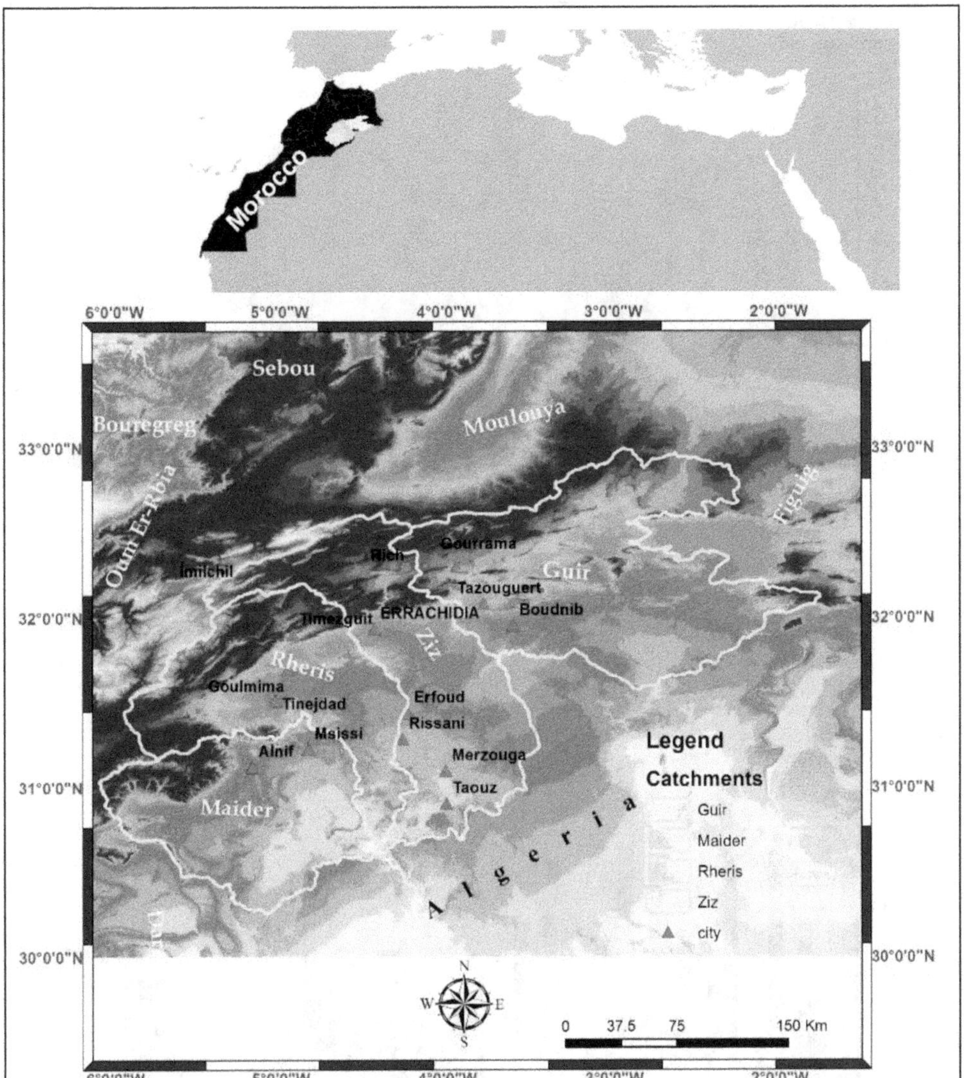

Figure 2. Average annual temperatures in Errachidia province
Source: (Ben Salem, 2014)

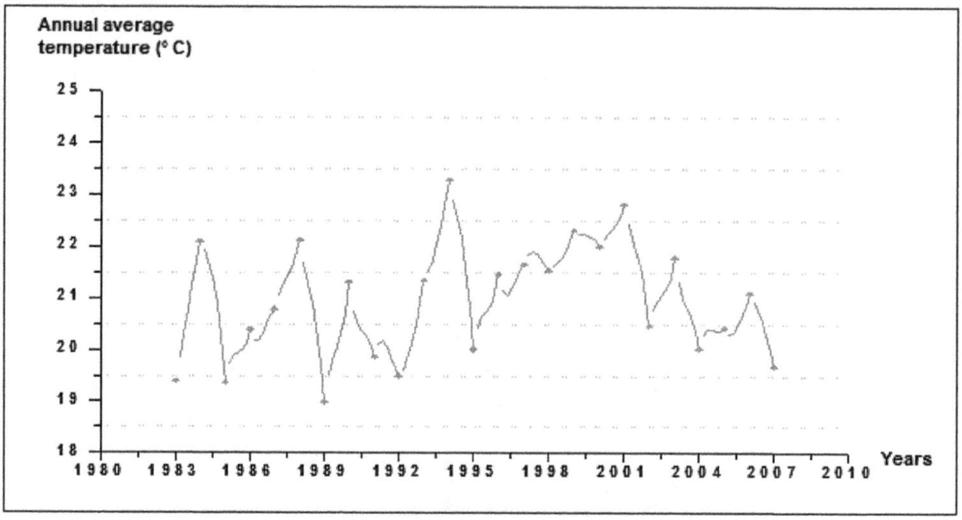

Figure 3. Average annual precipitation in Errachidia province
Source: (Ben Salem, 2014)

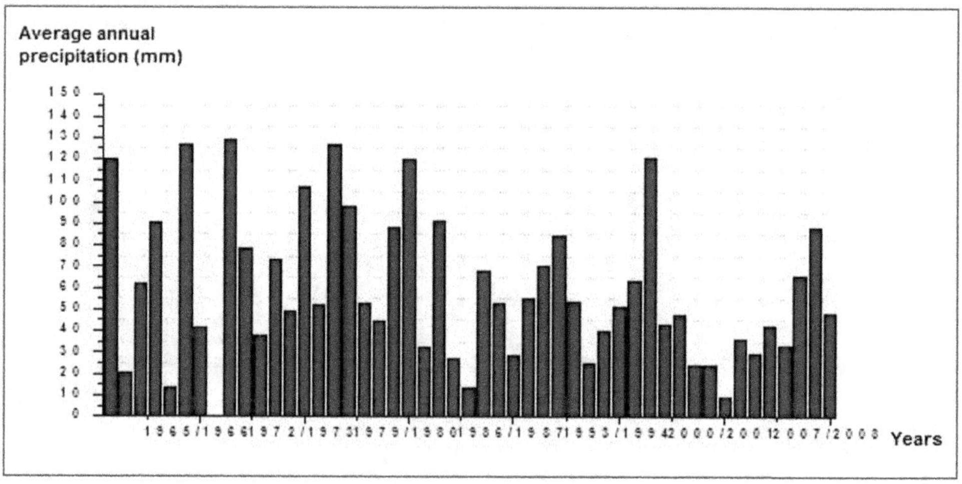

Table 1. Relationship between the aridity index, type of climate and vegetation

Value of I	Climate Type	Potential Vegetation Type
0 to 5	Hyperarid	Absolute desert
5 to 10	Arid	desert
10 to 20	Semi-arid	Steppe
20 to 30	Sub wet	Natural meadow, forest
30 to 40	wet	forest
40 to 55	Very wet	forest

Source: (Ait Ouhamchich, et al., 2018)

and vegetation index variables were assessed using the Pearson correlation coefficient (r). Differences are only considered significant if P.0.05.

Calculating the Aridity Index

According to the Word Reference Random House Unabridged Dictionary of American English, Aridity index refers to “a number indicating how much more precipitation could be lost by evapotranspiration if it were available than is lost at a given location”. The Martonne aridity index was used characterize the climate of the study area. This index corresponds to the annual precipitation height (P) and the average annual temperature (T) ratio (Eq1).

$$I = P\left(T + 10\right) \quad (1)$$

Where P, Annual rainfall height (in mm) and T, Annual temperature average (in °C) (Sen Roy & Balling, 2004).

De Martonne’s Index of Aridity is an empirical method that allows showing the transition from an area to another (Ait Ouhamchich, et al., 2018).This index reflects the climatic conditions using the low average annual rainfall and the strong deficit comparison with temperature. According to Ait Ouhamchich et al., 2018 “The high value of this index due to high rainfall and/or low temperatures led to a wet climate, and the opposite to dry climate as summarized in Table 1”.

Figure 4. Geographical repartition of cutaneous leishmaniasis in the study area in 2010
Source: (Authors)

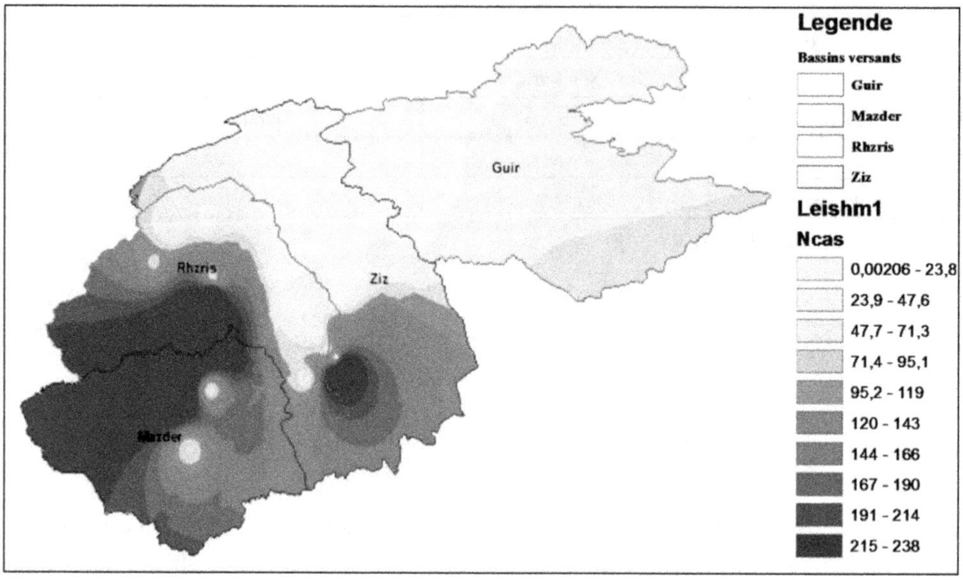

Normalized Difference Vegetation Index (NDVI)

The NDVI is a standardized index to generate an image showing plant cover (relative biomass). It is used worldwide to monitor drought, agricultural production, help prevent fires, and map desertification (Lillesand, Kiefer, & Chipman, 2004). The Vegetation Index data was downloaded from Moderate Resolution Imaging Spectroradiometer (MODIS) satellite imagery from January to December 2010 corresponding to the same period of climatic factors (http://reverb.echo.nasa.gov/reverb). The images were processed for the average NDVI calculated each month of 2010.

Results

The Spread of Leishmaniasis in Errachidia Province

Municipality Scale

The results (Figure 4) show a high concentration of cases in the municipalities occupying the western part of Errachidia province, while those in the north and east were less affected. The most affected commune was the rural municipality of Ferkla *Oulia* with a number of cases between 660 and 700. The rural commune

Figure 5. Distribution of cutaneous leishmaniasis cases by the type of area.
Source: (Authors)

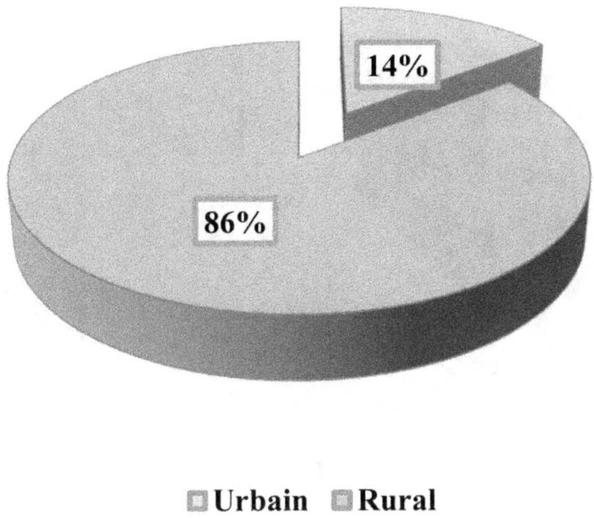

Sfalat recorded 250 to 400 cases, occupying the second place. However, negligible cases were noted in urban municipalities.

Depending on the Type of Environment

In the province of Errachidia, cutaneous leishmaniasis is practically common in rural and urban areas, but it has an uneven incidence. Indeed, it is more important in rural areas where 86% of cases are recorded compared to 14% in urban areas (Figure 5). This slight predominance of rural cases is observed in all years because of the habits and behaviors of the population, plus the presence of cracked stones in the walls of houses that serve as a place of reproduction or rest of sandflies. In addition to these conditions, lack of hygiene constitutes another impact factor.

The Distribution of Leishmaniasis by Gender

Figure 6 shows that CL affects both male and female and women are the most affected (57%), while men are less affected (43%).

Distribution of Leishmaniasis According to Screening

Two forms of surveillance – passive and active case-detection or searches – are mainly applied in disease control programs (WHO, 2019). 'Passive case detection' is detection triggered by patients seeking care for their illness from clinicians working in static health facilities. Clinicians who manage a case should notify it to

Figure 6. Percentage of cutaneous leishmaniasis by gender.
M: Male; F: Female
Source: (Authors)

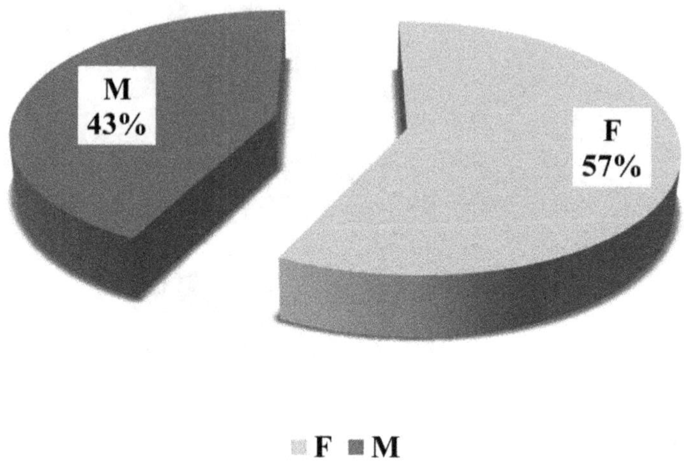

the appropriate epidemiological surveillance system, even in low or non-endemic countries, to obtain more accurate information of the leishmaniasis burden worldwide (WHO, 2019).

'Active case detection' or search means that health staff reaches out to the community and systematically screen the population to find cases of leishmaniasis. It helps to reduce disease transmission by shortening the infectious period of patients; and earlier diagnosis and treatment improves treatment outcomes of patients (WHO, 2019).

According to figure 7, the highest percentage appears in passive screening (65%), which shows that the population is aware of the dangerousness of the epidemic.

Distribution of Leishmaniasis Depending on the Year

According to the figures 8 and 9, more than 4407 cases were recorded during 2010, which is considered the most affected year in the last eight years. For this reason, it was used as a reference year.

Leishmaniasis is present all year round; however, there is a slight increase in cases during the rainy season compared to the dry season. December and January had the highest number of consultations (878 and 925) and those in May and June and July were the lowest (40, 19, and 17) (Figure 9).This can be explained by an incubation period that lasts about three months, so the first lesions appear after this period. Thus, the peaks observed in May, June, and July reflect the high levels of transmission that occurred in December and January.

Figure 7. The percentage of leishmaniasis depending on the type of screening. A, Active screening and P, passive screening
Source: (Authors)

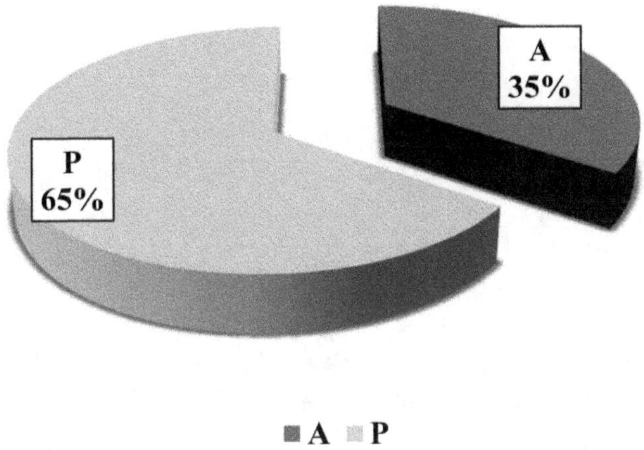

Figure 8. Leishmaniasis cases (Y axis) reportedbetween2010 and 2017 in the study area.
Source: (Authors)

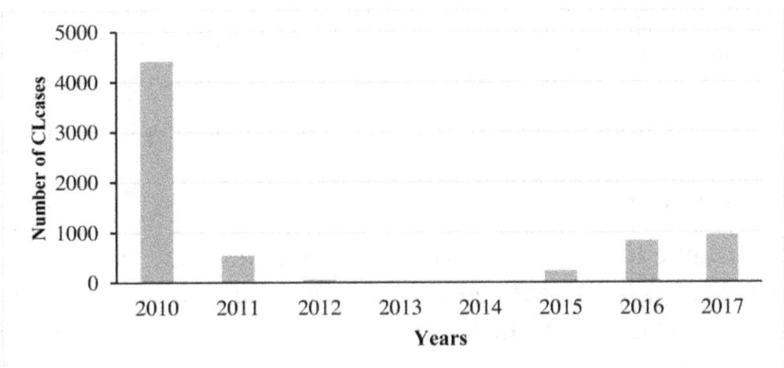

Impact of Climate Factors on the Spread of Leishmaniasis

The key factors responsible for the spatial-temporal evolution of leishmaniasis are currently unknown. It may be due to rodent population dynamics, vector dispersal, inter-annual climatic fluctuations, or /and high summer temperatures in some years. These last parameters promote the emergence of vectors of the disease. Other factors can also be considered like vegetation or cropping area (Karmaoui et al., 2019), soil

Figure 9. Monthly Leishmaniasis cases (Y axis) reportedbetween 2010 and 2012 inErrachidia
Source: (Authors)

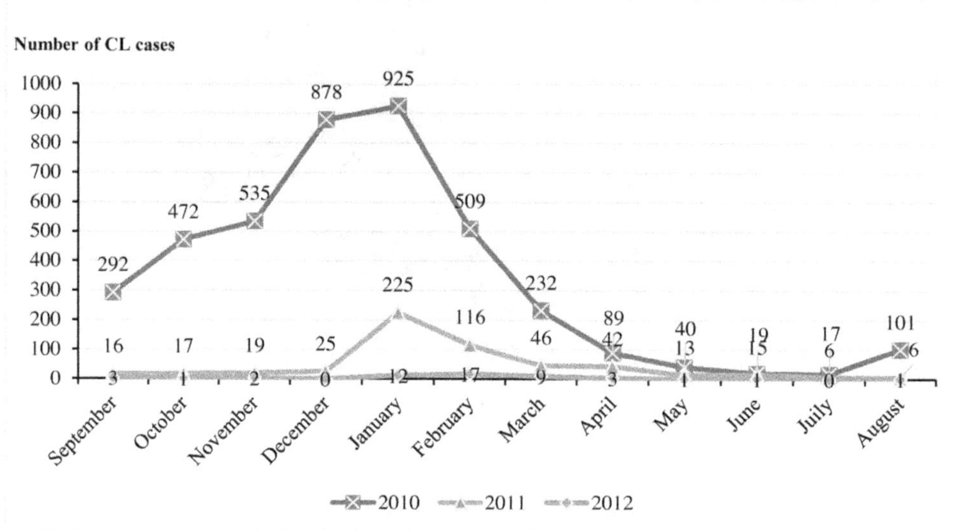

type, gradual increase in transmission in cities, displacement populations, deterioration of socio-economic conditions on the outskirts of cities, socio-cultural traits, and the evolution of biological diagnostic techniques (Rioux, 2007). The evaluation of the importance of these parameters will be critical for epidemiological understanding of the characteristics of leishmaniasis and can be used to predict the future evolution.

Temperature Effect

According to figure 10, the number of cases of leishmaniasis recorded in 2010 varied from a municipality to another, the very high number is recorded in *Farkla Oulia* (615 cases), while the lowest number was recorded in *Amellago* and in *Tadighouste*. The maximum temperature being regular the low temperature was recorded in the commune of Errachidia (28°C, 26°C), the highest temperature was recorded at *Ferkla Oulia* (32°C). So, there is a meaning between the number of cases and the maximum temperature. The municipality of Sidi Ayad, the number of cases of leishmaniasis recorded is very low, this is due to the unfavorable climatic conditions. This commune is characterized by the lowest recorded minimum temperature (6.6 ° C) and a low maximum temperature (30.02 ° C), similarly for the commune of Errachidia (9.6 °C and 28.03 °C) in which a small number of cases of leishmaniasis were recorded.

Figure 10. Evolutions of number of cases and temperature according to municipalities
Source: (Authors)

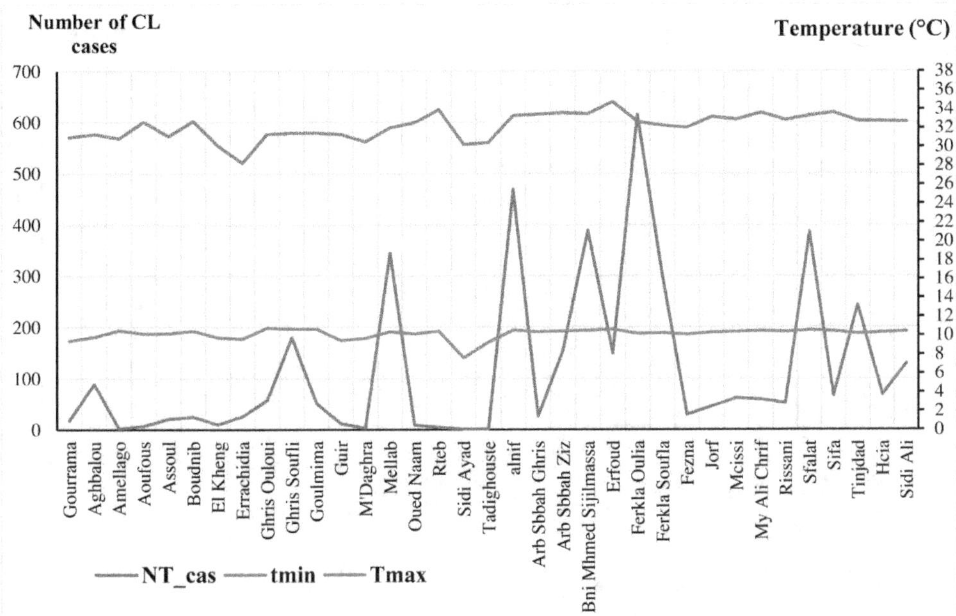

Aridity Index Effect

According to figure 11, there is a large irregularity of the aridity index in the various municipalities; the highest recorded aridity index and Rteb's with a score of 0.54, therefore the number of cases is negligible for the common Farkla Oulia which is recorded the largest number of cases the aridity index value is low 0.33.

From table 2 it can be concluded that the common differences of the aridity index is between 0 and 5, which gave a general idea about the type of climate that is hyper arid and also about the vegetation that is desert.

NDVI Index Effect

Figure 12 shows the relationship between NDVI and the evolution of the number of cases of leishmaniasis in Errachidia province during 2010.There is a great irregularity of NDVI between the different municipalities, max NDVI is registered in the communes of *Boudneb, Ghuis, Gruir,* and *M'daghra* with a score of 0.2 tends that the minimal NDVI is registered in the communes of *Alnif, Sfalat,* and *Hcia* with a score of 0.09. This low NDVI can reflect on the maximum number of cases of leishmaniasis recorded in the province.

Figure 11. Number of CL cases and the aridity index by municipality
Source: (Authors)

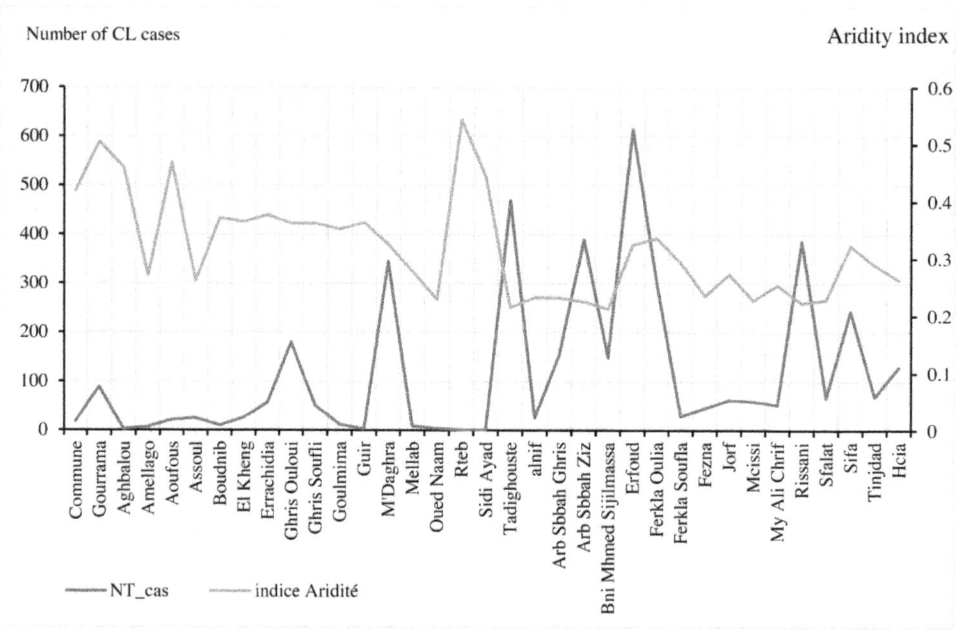

Figure 12. Number of CL cases and the NDVI vegetation index by municipality
Source: (Authors)

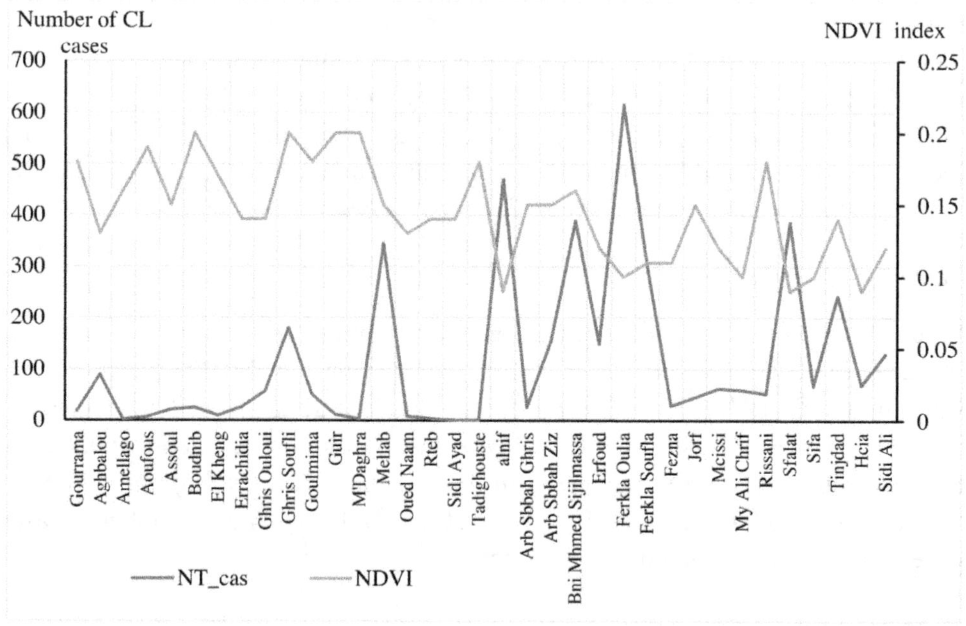

Table 2. Bilateral correlations between the number of CML cases and climatic conditions

		TN-Cases	Pt	NDVI	Tmin	Tmax	RH	WS	AI	Ev	Tm
TN-Cases	Pearson Correlation	1	-0,270	-0,416*	0,294	0,345*	-0,243	-0,105	-0,287	0,657**	0,364*
	Gis. (bilateral)		0,117	0,013	0,086	0,042	0,160	0,547	0,094	0,000	0,032
	N	35	35	35	35	35	35	35	35	35	35

*. The correlation is significant at the 0.05 (bilateral) level. The correlation is significant at level 0.01 (bilateral).

Abreviations in table 2: Total number of CL cases (TN-cases), Total annual precipitation (Pt), Normalized Difference Vegetation Index (NDVI),: minimum temperature (Tmin), maximum temperature (Tmax), Relative humidity *(RH), wind speed (WS), Aridity index (AI), Evapration (Ev), mean temperature (Tm), Sites number (N).*

Table 2 shows the results of bilateral correlations between the number of CML cases and climatic conditions. In most cases there is a significant correlation between minimum temperature, maximum temperature, precipitation, relative humidity, evaporation, wind speed and the distribution of the number of CML cases.

DISCUSSION

The results of this study should be of importance to understand the environmental risk factors associated with high prevalence and establishment of vector population. Consequently, this can help for better understanding of the transmission and guiding the target control programs in the region.

The number of cases, favored by agricultural development and the exploitation of new water resources (dams, culinary lakes, artesian wells, etc.) that have led to ecological changes creating wet microclimates. These conditions can be favorable to the development of the vector of the disease (Phlebotomus) and thus to the cycles of leishmaniasis. These ecological changes would also be responsible for the geographical extension of parasitosis.

In most cases, there is a significant correlation between minimum temperature, maximum temperature, precipitation, relative humidity, evaporation, wind speed and the distribution of the number of cases. Karmaoui et al., (2019) found a negative correlation between ZCL and temperature and an average correlation with precipitation. Bounoua et al., (2013) proposed that the increase in the reported ZCL cases is related to an increase in the minimum temperature, which allowed sand fly larvae to survive winters, thereby creating conditions suitable for an endemicity that did not previously exist. Greenness expressed in NDVI influences sand fly spatial repartition (Bounoua et al., 2013). In this study, NDVI showed negative correlation with case number of ZCL (Table 2) but is showed positive correlation with densities of *P. sergenti* males and females, but a decrease in density of both sexes of *P. papatasi* was observed with increase of NDVI values (Boussaa et al., 2016).The correlation of NDVI and sand fly density was only significant in case of *P. sergenti* females. The correlation of NDVI and fly density was only significant in case of *P. sergenti* females (Boussaa et al., 2016).

Leishmania Tropica (LT) in the Municipality of Assoul

According to the collected data, the *L. tropica* is registered only at the locality of *Agouray*, a rural commune of *Assoul*. In 2010, a new active focus of the disease transmission in *L. tropica* appeared at the town of *Agouray* commune of *Assoul* in the province of Errachidia.

This emergence of a new outbreak can be related to the climatic change experienced in the country, in the light of entomological surveys carried out along north-south transect routes perpendicular to the bioclimatic floors (since the cedar forests from the Rif to the pre-Saharan areas) showed that the sub-humid and semi-arid floors (especially upper and middle) would change floors and sub-stresses an evolution would double the area favorable to *L. tropica* (De la Rocque & Rioux, 2008).

CONCLUSION

Morocco, which is one of the most affected countries by leishmaniasis in North Africa. It is characterized by several bioclimatic levels ranging from the Mediterranean climate in the north to the Saharan climate in the south, through vast semi-arid areas. Because of its large rural population, it has a favorable ground for the spread of cutaneous and visceral forms of leishmaniasis. This would be related to climate changes. However, this phenomenon cannot be conclusively linked to global warming. Indeed, after analyzing epidemiological and meteorological data from the province of Errachidia, a real impact of climate change on this disease vector was found.

The results of this study suggest benchmarks of vector distribution and LC risk in Morocco. Climate factors (temperature, aridity, and humidity and wind speed) with the NDVI index can be considered important factors. Climate change plays an important role in populations of pathogen-carrying insects. However, for vector-borne diseases, anthropogenic factors (demography, development, urbanization) often play more important role. In fact, changes in practices (intra- and inter-continent movements, and trade), and socio-economic factors (degradation of health systems, weakening of surveillance systems) are significant examples.

There is a clear link between climatic factors and the incidence of cutaneous leishmaniasis, medium-term climate forecasts, and satellite monitoring of vegetation cover are essential tools for developing an alert system for cutaneous leishmaniasis.

REFERENCES

Ait Ouhamchich, K., Karaoui, I., Arioua, A., Kasmi, A., Elhamdouni, D., Elfiraoui, F., & Nabih, N. (2018). Climate Change Trend Observations in Morocco: Case Study of Beni Mellal-Khenifra and Darâa-Tafilalt Regions. *Journal of Geoscience and Environment Protection*, *6*(07), 34–50. doi:10.4236/gep.2018.67003

Ben Salem, A. (2014). *Vulnérabilité et adaptation aux changements climatiques dans les oasis du Tafilalet- Maroc*. Marrakech: Univeristé Cadi Ayyad.

Bounoua, L., Kahime, K., Houti, L., Blakey, T., Ebi, K., Zhang, P., ... Boumezzough, A. (2013). Linking climate to incidence of zoonotic cutaneous leishmaniasis (L. major) in pre-Saharan North Africa. *International Journal of Environmental Research and Public Health*, *10*(8), 3172–3191. doi:10.3390/ijerph10083172 PMID:23912199

Boussa, S. (2008). Épidémiologie des leishmanioses dans la région de Marrakech, Maroc: effet de l'urbanisation sur la répartition spatio-temporelle des phlébotomes et caractérisation moléculaire de leurs populations, thèse de doctorat en Ecologie-Epidémiologie, Université Louis Pasteur Strasbourg, France.

Boussaa, S., Kahime, K., Samy, A., Ben Salem, A., & Boumezzough, A. (2016). Species composition of sand flies and bionomics of Phlebotomus papatasi and P.sergenti (Diptera: Psychodidae) in cutaneous leishmaniasis endemic foci, Morocco. *Parasites & Vectors*, *9*(1), 60. doi:10.118613071-016-1343-6 PMID:26830689

De la Rocque, S., & Rioux, J. (2008). Influence des changements climatiques sur l'épidémiologie des maladies transmissibles. *Bulletin de la Société de Pathologie Exotique*, *101*(3), 213–219. PMID:18681214

Hufty, A. (2001). *Introduction à la climatologie: Le rayonnement et la température, l'atmosphère, l'eau, le climat et l'activité humaine*. Presses Université Laval.

Karmaoui, A., Zerouali, S., & Kahime, K. (2019). Climate Impact on Seasonal Distribution of Zoonotic Cutaneous Leishmaniasis in Southern Morocco. In *Handbook of Research on Global Environmental Changes and Human Health* (pp. 374–388). IGI Global. doi:10.4018/978-1-5225-7775-1.ch019

Karmaoui, A., Zerouali, S., Shah, A. A., Yacoubi-Khebiza, M., & El Qorchi, F. (2019). Ecosystem Services-Climate-Health Associations: Water-Climate-Leishmaniasis Nexus in an Endemic Focus of Zoonotic Cutaneous Leishmaniasis. In Climate Change and Its Impact on Ecosystem Services and Biodiversity in Arid and Semi-Arid Zones (pp. 280-290). IGI Global.

Lillesand, T., Kiefer, R., & Chipman, J. (2004). *Remote Sensing and Image Interpretation* (5th ed.). New York: John Wiley.

Mahjour, J., Akalay, O., & Saddiki, A. (1992). *Les leishmanioses au Maroc de l'analyse écoépidémiologie à la prévention*. DEPS Bul Epidém.

MS. (2019). Ministère de la santé Maroc. Récupéré sur https://www.sante.gov.ma/Reglementation/Pages/Accueil.aspx]. *MD Medical Newsmagazine*, *11*, 15.

Rguioui, I. (2006). Epidémiologie de la leishmaniose cutanée à SEFROU. Rabat: Thèse de doctorat en pharmacieFaculté de médecine et de pharmacie, Université Mohammed V-Rabat, Maroc.

Rioux, J. (2007). Le paradigme « écopathologie » Son application à l'épidémiologie des leishmanioses. Académie des sciences et lettres de Montpelier.

Sadqi, M. (2013). Profil épidémiologique des leishmanioses cutanées dans la province de Ouarzazate. Thèse de doctorat en pharmacie, Faculté de médecine et de pharmacie, Université Mohammed V-Souissi, Maroc.

Sen Roy, S., & Balling, R. (2004). Trends in Extreme Daily Precipitation Indices in India. *International Journal of Climatology*, *24*(4), 457–466. doi:10.1002/joc.995

WHO. (2019, Dec. 25). World Health Organization. Récupéré sur World Health Organization: https://www.who.int/leishmaniasis/surveillance/en/

Compilation of References

Abraham, D., Rotman, H. L., Haberstroh, H. F., Yutanawiboonchai, W., Brigandi, R. A., Leon, O., ... Schad, G. A. (1995). *Strongyloides stercoralis*: Protective immunity to third-stage larvae in BALB/cByJ mice. *Experimental Parasitology*, *80*(2), 297–307. doi:10.1006/expr.1995.1036 PMID:7895840

Adenusi, A. A. O., & Adenusi, A. O. (2003). Comparison of ivermectin and thiabendazole in the treatment of uncomplicated Strongyloides stercoralis infection. *African Journal of Biotechnology*, *2*(11), 465–469. doi:10.5897/AJB2003.000-1093

Ahern, M., Kovats, R. S., Wilkinson, P., Few, R., & Matthies, F. (2005). Global health impacts of floods: Epidemiologic evidence. *Epidemiologic Reviews*, *27*(1), 36–46. doi:10.1093/epirev/mxi004 PMID:15958425

Ait Ouhamchich, K., Karaoui, I., Arioua, A., Kasmi, A., Elhamdouni, D., Elfiraoui, F., & Nabih, N. (2018). Climate Change Trend Observations in Morocco: Case Study of Beni Mellal-Khenifra and Darâa-Tafilalt Regions. *Journal of Geoscience and Environment Protection*, *6*(07), 34–50. doi:10.4236/gep.2018.67003

Allen, S. J., O'donnell, A., Alexander, N. D. E., Alpers, M. P., Peto, T. E. A., Clegg, J. B., & Weatherall, D. J. (1997). α+-Thalassemia protects children against disease caused by other infections as well as malaria. *Proceedings of the National Academy of Sciences of the United States of America*, *94*(26), 14736–14741. doi:10.1073/pnas.94.26.14736 PMID:9405682

Alvarez, J. (1978). Strongyloidiasis en Guayaquil. Nuevas nociones epidemiologicas y biologicas. *Anales de Medicina y Cirugía*, *2*, 1–7.

Anderson, R. M. (1988). The role of mathematical models in the study of HIV transmission and the epidemiology of AIDS. *Journal of Acquired Immune Deficiency Syndromes*, *1*(3), 241–256. PMID:3216309

Arroyo, R., Troper, L., & Vargas, G. (1985). Estrongiloidiasis fatal: Studio de ocho casos de autopsia. *Patología (México)*, *23*, 135–146.

Atkinson, J. A., Vallely, A., Fitzgerald, L., Whittaker, M., & Tanner, M. (2011). The architecture and effect of participation: A systematic review of community participation for communicable disease control and elimination. Implications for malaria elimination. *Malaria Journal, 10*(1), 225. doi:10.1186/1475-2875-10-225 PMID:21816085

Attygalle, R. (1996). Contribution of home garden tree species towards family nutrition – case study, In H. P. M. Gunasena (Ed.), *Proceedings of the Forth National Workshop on Multipurpose Trees: Research and Development*, pp. 129–143, Kandy, Sri Lanka, April 1-3, 1994.

Bai, L., Ding, G., Gu, S., Bi, P., Su, B., Qin, D., ... Liu, Q. (2014). The effects of summer temperature and heat waves on heat-related illness in a coastal city of China, 2011–2013. *Environmental Research, 132*, 212–219. doi:10.1016/j.envres.2014.04.002 PMID:24815333

Ban, N., & Coomes, O. T. (2004). Home Gardens in Amazonian Peru: Diversity and Exchange of Planting Material. *Geographical Review, 94*(3), 348–367. doi:10.1111/j.1931-0846.2004.tb00177.x

Bannon, J. P., Fater, M., & Solit, R. (1995). Intestinal ileus secondary to Strongyloides stercoralis infection: Case report and review of the literature. *The American Surgeon, 61*, 377–380. PMID:7893110

Bañuelos, S., Martinez, M. V., Mitchell, C., & Prieto-Langarica, A. (2019). Using mathematical modelling to investigate the effect of the sexual behaviour of asymptomatic individuals and vector control measures on Zika. *Letters in Biomathematics, 6*(1), 1–19. doi:10.1080/23737867.2019.1624631

Barrett, K. E., Neva, F. A., Gam, A. A., Cicmanec, J., London, W. T., Phillips, J. M., & Metcalfe, D. D. (1988). The immune response to nematode parasites: modulation of mast cell numbers and function during Strongyloides stercoralis infections in nonhuman primates. *American Journal of Tropical Medicine and Hygiene 38*: 574-581.

Beggs, P. J. (2000). Impacts of climate and climate change on medications and human health. *Australian and New Zealand Journal of Public Health, 24*(6), 630–632. doi:10.1111/j.1467-842X.2000.tb00531.x PMID:11215015

Ben Salem, A. (2014). *Vulnérabilité et adaptation aux changements climatiques dans les oasis du Tafilalet- Maroc*. Marrakech: Univeristé Cadi Ayyad.

Bernoulli, D., & Blower, S. (2004). An attempt at a new analysis of the mortality caused by smallpox and of the advantages of inoculation to prevent it. *Reviews in Medical Virology, 14*(5), 275–288. doi:10.1002/rmv.443 PMID:15334536

Berthoud, F., & Berthoud, S. (1975). A propos de 18 cas d'anguillulose diagnostiques a Geneve. *Schweizerische Medizinische Wochenschrift, 105*, 1110–1115. PMID:1162316

Bisoffi, Z., Buonfrate, D., Sequi, M., Mejia, R., Cimino, R. O., Krolewiecki, A. J., ... Nutman, T. B. (2014). Diagnostic accuracy of five serologic tests for *Strongyloides stercoralis* infection. *PLoS Neglected Tropical Diseases, 8*(1). doi:10.1371/journal.pntd.0002640 PMID:24427320

Blaylock, J. R., & Gallo, A. E. (1983). Modeling the Decision to Produce Vegetables at Home. *Journal of Agricultural Economics*, *65*(4), 722–729. doi:10.2307/1240460

Boelee, E. (1999). Irrigation ecology of schistosomiasis: environmental control options in Morocco. (PhD thesis), Wageningen University, The Netherlands.

Boelee, E., & Laamrani, H. (2004). Environmental control of schistosomiasis through community participation in a Moroccan oasis. *Tropical Medicine & International Health*, *9*(9), 997–1004. doi:10.1111/j.1365-3156.2004.01301.x PMID:15361113

Boggild, A. K., Libman, M., Greenaway, C., & McCarthy, A. E. (2016). CATMAT statement on disseminated strongyloidiasis: Prevention, assessment and management guidelines. *Canada Communicable Disease Report*, *42*(1), 12–19. doi:10.14745/ccdr.v42i01a03 PMID:29769976

Bounoua, L., Kahime, K., Houti, L., Blakey, T., Ebi, K., Zhang, P., ... Boumezzough, A. (2013). Linking climate to incidence of zoonotic cutaneous leishmaniasis (L. major) in pre-Saharan North Africa. *International Journal of Environmental Research and Public Health*, *10*(8), 3172–3191. doi:10.3390/ijerph10083172 PMID:23912199

Boussa, S. (2008). Épidémiologie des leishmanioses dans la région de Marrakech, Maroc: effet de l'urbanisation sur la répartition spatio-temporelle des phlébotomes et caractérisation moléculaire de leurs populations, thèse de doctorat en Ecologie-Epidémiologie, Université Louis Pasteur Strasbourg, France.

Boussaa, S., Kahime, K., Samy, A., Ben Salem, A., & Boumezzough, A. (2016). Species composition of sand flies and bionomics of Phlebotomus papatasi and P.sergenti (Diptera: Psychodidae) in cutaneous leishmaniasis endemic foci, Morocco. *Parasites & Vectors*, *9*(1), 60. doi:10.118613071-016-1343-6 PMID:26830689

Breitenfeld, V., & Iskra, M. (1984). Strongiloidoza u nas i u svijetu. *Lijecnicki Vjesnik*, *106*, 33–38. PMID:6708716

Brindley, P. J., Gam, A. A., Pearce, E. J., Poindexter, R. W., & Neva, F. A. (1988). Antigens from the surface and excretions/secretions of the filariform larva of *Strongyloides stercoralis*. *Molecular and Biochemical Parasitology*, *28*(17), 1–180. PMID:3386679

Brumpt, Le., Sang, H. T., & Jaeger, G. (1972). Quelques reflexions apropos du parasitisme sanguin et intestinal dans deux villages d'Afrique Centrale. *Bulletin de la Société de Pathologie Exotique*, *65*, 263–270.

Bruun, B., & Aagaard-Hansen, J. (2008). *The social context of schistosomiasis and its control: an introduction and annotated bibliography*. World Health Organization.

Buonfrate, D., Sequi, M., Mejia, R., Cimino, R. O., Krolewiecki, A. J., Albonico, M., ... Bisoffi, Z. (2015). Accuracy of five serologic tests for the follow up of Strongyloides stercoralis infection. *PLoS Neglected Tropical Diseases*, *9*(2). doi:10.1371/journal.pntd.0003491 PMID:25668740

Bustinza, R., Lebel, G., Gosselin, P., Bélanger, D., & Chebana, F. (2013). Health impacts of the July 2010 heat wave in Quebec, Canada. *BMC Public Health, 13*(1), 56. doi:10.1186/1471-2458-13-56 PMID:23336593

Cairncross, S., Shordt, K., Zacharia, S., & Govindan, B. K. (2005). What causes sustainable changes in hygiene behaviour? A cross-sectional study from Kerala, India. *Social science & medicine (1982), 61*(10), 2212–2220.

Capell, F. S., Pujol, F. R., Garau, A. J., Pallares, G. R., & Campo, G. E. (1982). Hiperinfestacion por *Strongyloides stercoralis. Medicina Clínica, 79*, 232–235. PMID:6983009

Carvalho, E. (1978). Strongyloidiasis. *Clinics in Gastroenterology, 7*, 179–200. PMID:627098

Centers for Disease Control and Prevention. (2000). Infectious Diseases Society of America, American Society of Blood and Marrow Transplantation. Guidelines for preventing opportunistic infections among hematopoietic stem cell transplant recipients. *MMWR. Recommendations and Reports, 49*(RR-10), 1–125.

Cerutti, B., Tereanu, C., Domenighetti, G., Cantoni, E., Gaia, M., Bolgiani, I., ... Cassis, I. (2006). Temperature related mortality and ambulance service interventions during the heat waves of 2003 in Ticino (Switzerland). *Sozial- und Praventivmedizin, 51*(4), 185–193. doi:10.100700038-006-0026-z PMID:17193780

Chang, F. C., & Wallace, J. M. (1987). Meteorological conditions during heat waves and droughts in the United States Great Plains. *Monthly Weather Review, 115*(7), 1253–1269. doi:10.1175/1520-0493(1987)115<1253:MCDHWA>2.0.CO;2

Chaudhry, Q. Z., Rasul, G., Kamal, A., Mangrio, M. A., & Mahmood, S. (2015). Task Force/ Fact Finding Mission by GOP on "Causes of Severe Heat Wave in Karachi. Technical Report, 2015. Coley, D.

Chaudhury, S. K., Gore, J. M., & Ray, K. S. (2000). Impact of heat waves over India. *Current Science, 79*(2), 153–155.

Cheesbrough, M. (1987). *Medical laboratory manual for tropical countries* (2nd ed., Vol. 1). Cambridge: Tropical Health Technology and Butterworths.

Chen, J. J., Lee, C. M., & Changchan, C. S. (1994). Duodenal *Strongyloides stercoralis* infection. *Endoscopy, 26*(02), 272. doi:10.1055-2007-1008965 PMID:8026386

Chiodini, P. L., Reid, A. J., Wiselka, M. J., Firmin, R., & Foweraker, J. (2000). Parenteral ivermectin in Strongyloides hyperinfection. *Lancet, 355*(9197), 43–44. doi:10.1016/S0140-6736(99)02744-0 PMID:10615895

Chitnis, N., Schapira, A., Schindler, C., Penny, M. A., & Smith, T. A. (2018). Mathematical analysis to prioritise strategies for malaria elimination. *Journal of Theoretical Biology, 45*(5), 118–130. doi:10.1016/j.jtbi.2018.07.007 PMID:30006002

Choi, J., & Meentemeyer, V. (2002). Climatology of persistent positive temperature anomalies for the contiguous United States (1950–1995). *Physical Geography, 23*(3), 175–195. doi:10.2747/0272-3646.23.3.175

Chowell, G., Mizumoto, K., Banda, J. M., Poccia, S., & Perrings, C. (2019). Assessing the potential impact of vector-borne disease transmission following heavy rainfall events: A mathematical framework. *Philosophical Transactions of the Royal Society of London. Series B, Biological Sciences, 374*(7), 1–9. PMID:31056044

Chu, C. (1994). *Community participation in public health: Definitions and conceptual framework. s, Ecological Public Health: From Vision to Practice.* Brisbane: Institute of Applied Environmental Research.

Ciftcioglu, G. C. (2017). Social preference-based valuation of the links between home gardens, ecosystem services, and human wellbeing in Lefke Region of North Cyprus. *Ecosystem Services, 25*, 227–236. doi:10.1016/j.ecoser.2017.05.002

Conway, D. J., Bailey, J. W., Lindo, J. F., Robinson, R. D., Bundy, D. A. P., & Bianco, A. E. (1993). Serum IgG reactivity with 41-, 31-, and 28-kDa larval proteins of *Strongyloides stercoralis* in individuals with strongyloidiasis. *The Journal of Infectious Diseases, 168*(3), 784–787. doi:10.1093/infdis/168.3.784 PMID:8354924

Conway, D. J., Lindo, J. F., Robinson, R. D., Bundy, D. A. P., & Bianco, A. E. (1994). *Strongyloides stercoralis*: Characterization of immunodiagnostic larval antigens. *Experimental Parasitology, 79*(2), 99–105. doi:10.1006/expr.1994.1069 PMID:8056086

Crompton, W. T., & Tulley, J. J. (1987). How much ascariasis is there in Africa? *Parasitology Today (Personal Ed.), 3*(4), 123–127. doi:10.1016/0169-4758(87)90054-8 PMID:15462932

Cromwell, E., Cooper, D., & Mulvany, P. (1991) Agricultural Biodiversity and Livelihoods: Issues and Entry Points for Development Agencies.

Cruz, R. V., Harasawa, H., Lal, M., Wu, S., Anokhin, Y., Punsalmaa, B., . . . Huu Nin, N. (2007). Asia. In Climate Change 2007: Impacts, Adaptation and Vulnerability. Contribution of Working Group II to the Fourth Assessment Report of the Intergovernmental Panel on Climate Change.

Czachor, J. S., & Jonas, A. P. (2000). Transmission of *Strongyloides steracolis* person to person. *Journal of Travel Medicine, 7*(4), 211–212. doi:10.2310/7060.2000.00063 PMID:11003736

D. P. E. La Direction Provinciale de l'Équipement (*DPE*) de TATA (Poste D'AKKA) Service Eau.

Dalip, J., Phillips, G. A., Jelinek, G. A., & Weiland, T. J. (2015). Can the elderly handle the heat? A retrospective case-control study of the impact of heat waves on older patients attending an inner-city Australian emergency department. *Asia-Pacific Journal of Public Health, 27*(2), NP1837–NP1846. doi:10.1177/1010539512466428 PMID:23212171

Datry, A., Hilmarsdottir, I., Mayorga-Sagastume, R., Lyagoubi, M., Gaxotte, P., Biligui, S., ... Gentilini, M. (1994). Treatment of *Strongyloides stercoralis* infection with ivermectin compared with albendazole: Results of an open study of 60 cases. *Transactions of the Royal Society of Tropical Medicine and Hygiene*, *88*(3), 344–345. doi:10.1016/0035-9203(94)90110-4 PMID:7974685

Daulagala, C., Weerahewa, J., Marambe, B., Pushpakumara, G., Silva, P., Punyawardena, R., ... Jana, S. (2013). Socio-economic characteristics of farmers influencing adaptation to climate change: Empirical results from selected homegardens in South Asia with emphasis on commercial orientation. *Sri Lanka Journal of Advanced Social Studies*, *2*(2). doi:10.4038ljass.v2i2.6208

Davis, J. R., Brownson, R. C., & Garcia, R. (1992). Family pesticide use in the home, garden, orchard, and yard. *Archives of Environmental Contamination and Toxicology*, *22*(3), 260–266. doi:10.1007/BF00212083 PMID:1616309

Dawn. (2019, March 2). *No TitHunger, disease "alarmingly high" in drought-hit Pakistan: Red Crossle*. Retrieved from https://www.dawn.com/news/1467107/hunger-disease-alarmingly-high-in-drought-hit-pakistan-red-cross

De la Rocque, S., & Rioux, J. (2008). Influence des changements climatiques sur l'épidémiologie des maladies transmissibles. *Bulletin de la Société de Pathologie Exotique*, *101*(3), 213–219. PMID:18681214

De Zoysa, M., & Inoue, M. (2014). Climate change impacts, agroforestry adaptation and policy environment in Sri Lanka. *Open Journal of Forestry*, *4*(5), 439–456. doi:10.4236/ojf.2014.45049

Delarocque-Astagneau, E., Hadengue, A., Degott, C., Vilgrain, V., Erlinger, S., & Benhamou, J. P. (1994). Biliary obstruction resulting from *Strongyloides stercoralis* infection. *Report of a case. Gut*, *35*(5), 705–706. doi:10.1136/gut.35.5.705 PMID:8200571

DeVault, G. A., King, J. W. Jr, Rohr, M. S., Landreneau, M. D., Brown, S. T. III, & McDonald, J. C. (1990). Opportunistic infections with *Strongyloides stercoralis* in renal transplantation. *Reviews of Infectious Diseases*, *12*(4), 653–671. doi:10.1093/clinids/12.4.653 PMID:2201067

Direction de la santé publique. (1940). BULLETIN DE L'INSTITUT D'HYGIÈNE DU MAROC.

Doumenge, J. P., Mott, K. E., Cheung, C., Villenave, D., Chapuis, O., Perrin, M. F., & Reaud-Thomas, G. (1987). Atlas of the global distribution of schistosomiasis. World Health Organization, Parasitic Diseases Programme. Presses Universitaires, Bordeaux, France.

Dreyer, G., Fernandes-Silva, E., Alves, S., Rocha, A., Albuquerque, R., & Addiss, D. (1996). Patterns of detection of *Strongyloides stercoralis* in stool specimens: Implications for diagnosis and clinical trials. *Journal of Clinical Microbiology*, *34*(10), 2569–2571. doi:10.1128/JCM.34.10.2569-2571.1996 PMID:8880521

Duggan, M. B. (2003). Nutritional update: Relevance to maternal and child health in East Africa. *African Health Sciences*, *3*(3), 136–143. PMID:14676720

Ejezie, G. C., & Otigbuo, I. N. (1987). The stabilising factors of parasitic infections in Nigeria. *Epidemiologie, Mikrobiologie, Imunologie*, *31*, 45–51. PMID:2883231

Ekvall, H. (2003). Malaria and anemia. *Current Opinion in Hematology, 10*(2), 108–114. doi:10.1097/00062752-200303000-00002 PMID:12579035

Eubank, S., Guclu, H., Kumar, V. A., Marathe, M. V., Srinivasan, A., Toroczkai, Z., & Wang, N. (2004). Modelling disease outbreaks in realistic urban social networks. *Nature, 429*(6), 180–184. doi:10.1038/nature02541 PMID:15141212

Ezzati, M., Lopez, A. D., Rodgers, A. A., & Murray, C. J. (2004). *Comparative quantification of health risks: the global and regional burden of disease attributable to selected major risk factors*. World Health Organization.

Faria, J. A. S. (1972). Prevalencia de helmintos em escolares de 7-15 anos na cidade de Salvador. *Revista da Sociedade Brasileira de Medicina Tropical, 6*(5), 261–264. doi:10.1590/S0037-86821972000500004

Fiillebom, F. H. (1926). Autoinfektion bei Strongyloidestriigem. *Arch Schiffs Tropenhyg, 30*, 721–732.

Forrer, A., Khieu, V., Schar, F., Hattendorf, J., Marti, H., Neumayr, A., ... Odermatt, P. (2017). *Strongyloides stercoralis* is associated with significant morbidity in rural Cambodia, including stunting in children. *PLoS Neglected Tropical Diseases, 11*(10). doi:10.1371/journal.pntd.0005685 PMID:29059195

Fouillet, A., Rey, G., Laurent, F., Pavillon, G., Bellec, S., Guihenneuc-Jouyaux, C., ... Hémon, D. (2006). Excess mortality related to the August 2003 heat wave in France. *International Archives of Occupational and Environmental Health, 80*(1), 16–24. doi:10.100700420-006-0089-4 PMID:16523319

Gann, P. H., Neva, F. A., & Gam, A. A. (1994). A randomized trial of single- and two-dose ivermectin versus thiabendazole for treatment of strongyloidiasis. *The Journal of Infectious Diseases, 169*(5), 1076–1079. doi:10.1093/infdis/169.5.1076 PMID:8169394

García Cruz, L. M., Gonzalez Azpeitia, G., Reyes Suarez, D., Santana Rodríguez, A., Loro Ferrer, J. F., & Serra-Majem, L. (2017). Factors associated with stunting among children aged 0 to 59 months from the central region of Mozambique. *Nutrients, 9*(5), 491. doi:10.3390/nu9050491 PMID:28498315

Genta, R. M. (1988). Predictive value of an enzyme-linked immunosorbent assay (ELISA) for the serodiagnosis of strongyloidiasis. *American Journal of Clinical Pathology, 89*(3), 391–394. doi:10.1093/ajcp/89.3.391 PMID:3348175

Genta, R. M. (1989). Global prevalence of strongyloidiasis: Critical review with epidemiologic insights into the prevention of disseminated disease. *Reviews of Infectious Diseases, 11*(5), 755–767. doi:10.1093/clinids/11.5.755 PMID:2682948

Genta, R. M., Gatti, S., Linke, M. J., Cevini, C., & Scaglia, M. (1988). Endemic strongyloidiasis in northern Italy: Clinical and immunological aspects. *The Quarterly Journal of Medicine, 258*, 679–690. PMID:3256899

Genta, R. M., & Lillibridge, J. P. (1989). Prominence of IgG4 antibodies in the human responses to *Strongyloides stercoralis* infection. *The Journal of Infectious Diseases*, *160*(4), 692–699. doi:10.1093/infdis/160.4.692 PMID:2794561

Gervas, H. E., Opoku, N. K. D. O., & Ibrahim, S. (2018). Mathematical Modelling of Human African Trypanosomiasis Using Control Measures. *Computational and Mathematical Methods in Medicine*, *2018*(5), 1–13. doi:10.1155/2018/5293568 PMID:30595713

Ghoshal, U. C., Alexender, G., Ghoshal, U., Tripathi, S., & Krishnani, N. (2006). *Strongyloides stercoralis* infestation in a patient with severe ulcerative colitis. *Indian Journal of Medical Sciences*, *60*(3), 106–110. doi:10.4103/0019-5359.22761 PMID:16569924

Giannoulis, E., Arvanitakis, C., Zaphiropoulos, A., Nakos, V., Karkavelas, G., & Haralambidis, S. (1986). Disseminated strongyloidiasis with uncommon manifestations in Greece. *The Journal of Tropical Medicine and Hygiene*, *89*, 171–178. PMID:3783811

Githeko, A. K., Lindsay, S. W., Confalonieri, U. E., & Patz, J. A. (2000). Climate change and vector-borne diseases: A regional analysis. *Bulletin of the World Health Organization*, *78*(4), 1136–1147. PMID:11019462

Government of Pakistan. (2017). *Annual Flood Report 2017*.

Greenwood, M. (1916). The application of mathematics to epidemiology. *Nature*, *97*(4), 243–244. doi:10.1038/097243a0

Grove, D. I. (1982). Treatment of strongyloidiasis with thiabendazole: An analysis of toxicity and effectiveness. *Transactions of the Royal Society of Tropical Medicine and Hygiene*, *76*(1), 114–118. doi:10.1016/0035-9203(82)90034-7 PMID:7080143

Grum, R. H. (2011). The central European and Russian heat event of July–August 2010. *Bulletin of the American Meteorological Society*, *92*(10), 1285–1286. doi:10.1175/2011BAMS3174.1

Gupta, S., Jain, A., Fanning, T. V., Couriel, D. R., Jimenez, C. A., & Eapen, G. A. (2006). An unusual cause of alveolar hemorrhage post hematopoietic stem cell transplantation: A case report. *BMC Cancer*, *6*(1), 87. doi:10.1186/1471-2407-6-87 PMID:16603072

Gustavo, S. J. (1993). Movement of the rhabditiform larva of *Strongyloides stercoralis*. *Lancet*, *342*, 13–10. PMID:8100289

Haines, A., Kovats, R. S., Campbell-Lendrum, D., & Corvalán, C. (2006). Climate change and human health: Impacts, vulnerability, and mitigation. *Lancet*, *367*(9528), 2101–2109. doi:10.1016/S0140-6736(06)68933-2 PMID:16798393

Hales, S., Edwards, S. J., & Kovats, R. S. (2003). Impacts on health of climate extremes. *Climate change and human health: Risks and responses*, 79-96.

HCP. (2004). HautCommissariat au Plan du Maroc, Recensement général de la population et de l'habitat 2004.

Henriquez-Camacho, C., Gotuzzo, E., Echevarria, J., White Jr, A. C., Terashima, A., Samalvides, F., ... & Plana, M. N. (2016). Ivermectin versus albendazole or thiabendazole for *Strongyloides stercoralis* infection. *Cochrane Database of Systematic Reviews*, (1). PMID:26778150

Hernández-Díaz, S., Peterson, K. E., Dixit, S., Hernández, B., Parra, S., Barquera, S., ... Rivera, J. A. (1999). Association of maternal short stature with stunting in Mexican children: Common genes vs. common environment. *European Journal of Clinical Nutrition*, *53*(12), 938–945. doi:10.1038j.ejcn.1600876 PMID:10602351

Hess, J. J., Malilay, J. N., & Parkinson, A. J. (2008). Climate change: The importance of place. *American Journal of Preventive Medicine*, *35*(5), 468–478. doi:10.1016/j.amepre.2008.08.024 PMID:18929973

Hethcote, H. W. (2000). The mathematics of infectious diseases. *SIAM Review*, *42*(4), 599–653. doi:10.1137/S0036144500371907

Hippner, P., Sumner, T., Houben, R. M., Cardenas, V., Vassall, A., Bozzani, F., & White, R. G. (2019). Application of provincial data in mathematical modelling to inform sub-national tuberculosis program decision-making in South Africa. *PLoS One*, *14*(1), 1–11. doi:10.1371/journal.pone.0209320 PMID:30682028

Hitinayake, H. M. G. S. B., De Costa, W. A. J. M., & Jayaweera, K. G. D. (1996). Food trees in multi-layered home gardens in different Agroecological Zones of Kandy District, In H. P. M. a Gunasen (Ed.), *Proceedings of the Seventh National Workshop on Multipurpose Trees: Multipurpose Trees for Food Security*, pp. 252–264. Kandy, Sri Lanka, November 5, 1999.

Hoermann, B., Banerjee, S., & Kollmair, M. (2010). *Labour migration for development in the western Hindu Kush-Himalayas: Understanding a livelihood strategy in the context of socioeconomic and environmental change*. Kathmandu: International Centre for Integrated Mountain Development; Available at http://lib.icimod.org/record/8050/file s/attachment_695.pdf

Holland, C. V., Crompton, D. W. T., Taren, D. L., Nesheim, M. C., Sanjur, D., Barbeau, I., & Tucker, K. (1987). Ascaris lumbricoides infection in pre-school children from Chiriqui Province, Panama. *Parasitology*, *95*(3), 615–622. doi:10.1017/S0031182000058030 PMID:3696782

Hotez, P. J., & Kamath, A., & Cappello, M. (Eds.). (2009). Neglected Tropical Diseases in Sub-Saharan Africa: Review of Their Prevalence, Distribution, and Disease Burden. PLoS Negl Trop Dis. 3(8). doi:. doi:10.1371/journal.pntd.0000412

Hotez, P. (2012). Fighting neglected tropical diseases in the southern United States. *BMJ (Clinical Research Ed.)*, *345*(sep13 1). doi:10.1136/bmj.e6112 PMID:22977143

Hotez, P. J. (2013, November). NTDs V.2.0: "blue marble health"—neglected tropical disease control and elimination in a shifting health policy landscape. *PLoS Neglected Tropical Diseases*, *7*(11). doi:10.1371/journal.pntd.0002570 PMID:24278496

Hoy, W. E., Roberts, N. J., Bryson, M. F., Bowles, C., Lee, J. C., Rivero, A. J., & Ritterson, A. L. (1981). Transmission of strongyloidiasis by kidney transplant? Disseminated strongyloidiasis in both recipients of kidney allografts from a single cadaver donor. *Journal of the American Medical Association, 246*(17), 1937–1939. doi:10.1001/jama.1981.03320170049031 PMID:7026811

Hufty, A. (2001). *Introduction à la climatologie: Le rayonnement et la température, l'atmosphère, l'eau, le climat et l'activité humaine*. Presses Université Laval.

Hunt, B. G. (2007). A climatology of heat waves from a multimillennial simulation. *Journal of Climate, 20*(15), 3802–3821. doi:10.1175/JCLI4224.1

Igra-Siegman, Y., Kapila, R., Sen, P., Kaminski, Z. C., & Louria, D. B. (1981). Syndrome of hyperinfection with *Strongyloides stercoralis*. *Reviews of Infectious Diseases, 3*(3), 397–407. doi:10.1093/clinids/3.3.397 PMID:7025145

Im, E. S., Pal, J. S., & Eltahir, E. A. (2017). Deadly heat waves projected in the densely populated agricultural regions of South Asia. *Science Advances, 3*(8). doi:10.1126ciadv.1603322 PMID:28782036

IPCC. (2007). Climate change 2007: Synthesis report. Geneva, World Meteorological Organization (WMO), United Nations Environmental Program (UNEP).

IPCC. (2007a). [*The physical science basis*. Geneva: Intergovernmental Panel on Climate Change. References]. *Climatic Change*, 2007.

Jana, S. K., & Roy, J. (2020). Biodiversity and Impacts of Climate Change in Home Gardens: Evidence From a Study in West Bengal, India. In Current State and Future Impacts of Climate Change on Biodiversity (pp. 113-134). IGI Global.

Jana, S., Roy, J., Marambe, B., Weerahewa, J., Pushpakumara, G., Silva, P., ... Premalal, S. (2015). Home gardens in the Paschim Medinipur District of West Bengal in India: A land use system with multiple benefits. *International Journal of Environment and Sustainable Development, 14*(2), 191–206. doi:10.1504/IJESD.2015.068606

Jongman, B., Winsemius, H. C., Aerts, J. C., de Perez, E. C., van Aalst, M. K., Kron, W., & Ward, P. J. (2015). Declining vulnerability to river floods and the global benefits of adaptation. *Proceedings of the National Academy of Sciences of the United States of America, 112*(18), E2271–E2280.

Junod, C. (1987). Etude retrospective de 1,934 cas de strongyloidose diagnostiques aParis (1970-1986). Origine geographique, Epidemiologie. *Bulletin de la Société de Pathologie Exotique, 80*, 357–369. PMID:3621398

Kalinda, C., Chimbari, M. J., & Mukaratirwa, S. (2017). Effect of temperature on the Bulinusglobosus—Schistosomahaematobium system. *Infectious Diseases of Poverty, 6*(1), 57. doi:10.118640249-017-0260-z PMID:28457230

Kamiya, Y., Nomura, M., Ogino, H., Yoshikawa, K., Siengsounthone, L., & Xangsayarath, P. (2018). Mothers' autonomy and childhood stunting: Evidence from semi-urban communities in Lao PDR. *BMC Women's Health, 18*(1), 70. doi:10.118612905-018-0567-3 PMID:29789003

Karmaoui, A., Zerouali, S., Shah, A. A., Yacoubi-Khebiza, M., & El Qorchi, F. (2019). Ecosystem Services-Climate-Health Associations: Water-Climate-Leishmaniasis Nexus in an Endemic Focus of Zoonotic Cutaneous Leishmaniasis. In Climate Change and Its Impact on Ecosystem Services and Biodiversity in Arid and Semi-Arid Zones (pp. 280-290). IGI Global.

Karmaoui, A. (2018). The cutaneous leishmaniasis vulnerability index (CLVI). *Acta Ecologica Sinica, 38*(4), 288–295. doi:10.1016/j.chnaes.2018.01.001

Karmaoui, A., Zerouali, S., & Kahime, K. (2019). Climate Impact on Seasonal Distribution of Zoonotic Cutaneous Leishmaniasis in Southern Morocco. In *Handbook of Research on Global Environmental Changes and Human Health* (pp. 374–388). IGI Global. doi:10.4018/978-1-5225-7775-1.ch019

Keeling, M. J., & Eames, K. T. (2005). Networks and epidemic models. *Journal of the Royal Society, Interface, 2*(4), 295–307. doi:10.1098/rsif.2005.0051 PMID:16849187

Kennedy, S., Campbell, R. M., Lawrence, J. E., Nichol, G. M., & Rao, D. M. (1989). A case of severe *Strongyloides stercoralis* infection with jejunal perforation in an Australian ex-prisoner-of-war. *The Medical Journal of Australia, 150*(2), 92–93. doi:10.5694/j.1326-5377.1989.tb136371.x PMID:2542737

Kermack, W. O., & McKendrick, A. G. (1927). A contribution to the mathematical theory of epidemics. *Proceedings of the Royal Society of London. Series A, Containing Papers of a Mathematical and Physical Character, 115*(772), 700–721. doi:10.1098/rspa.1927.0118

Kermack, W. O., & McKendrick, A. G. (1932). Contributions to the mathematical theory of epidemics. II.—The problem of endemicity. *Proceedings of the Royal Society of London. Series A, Containing Papers of a Mathematical and Physical Character, 138*(834), 55–83. doi:10.1098/rspa.1932.0171

Kermack, W. O., & McKendrick, A. G. (1933). Contributions to the mathematical theory of epidemics. III.—Further studies of the problem of endemicity. In *Proceedings of the Royal Society of London. Series A, Containing Papers of a Mathematical and Physical Character, 141*(843), 94–122. doi:10.1098/rspa.1933.0106

Keyantash, J., & Dracup, A. J. (2002). The Quantification of Drought: An Evaluation of Drought Indices. Bulletin of the American Meteorological Society. 83. 1167-1180. <1191: TQODAE>2.3.CO;2Dawn. (2019, March 2). *No TitHunger, disease "alarmingly high" in drought-hit Pakistan: Red Crossle*. Retrieved from https://www.dawn.com/news/1467107/hunger-disease-alarmingly-high-in-drought-hit-pakistan-red-cross doi:10.1175/1520-0477(2002)083

Khallaayoune, K., & Laamrani, H. (1992). Seasonal patterns in the transmission of Schistosoma haematobium in Attaouia, Morocco. *Journal of Helminthology, 66*(2), 89–95. doi:10.1017/S0022149X00012645 PMID:1640092

Khieu, V., Schar, F., Marti, H., Sayasone, S., Duong, S., Muth, S., & Odermatt, P. (2013). Diagnosis, treatment and risk factors of *Strongyloides stercoralis* in schoolchildren in Cambodia. *PLoS Neglected Tropical Diseases*, *7*(2). doi:10.1371/journal.pntd.0002035 PMID:23409200

Klinenberg, E. (2015). *Heat wave: A social autopsy of disaster in Chicago*. University of Chicago Press. doi:10.7208/chicago/9780226276212.001.0001

Kreyszig, E. (1978). *Introductory functional analysis with applications*. New York: Wiley.

Krolewiecki, A. J., Lammie, P., Jacobson, J., Gabrielli, A.-F., Levecke, B., Socias, E., ... Albonico, M. (2013). A public health response against *Strongyloides stercoralis*: Time to look at soil-transmitted helminthiasis in full. *PLoS Neglected Tropical Diseases*, *7*(5), 21–65. doi:10.1371/journal.pntd.0002165 PMID:23675541

Laamrani, H., Khallaayoune, K., Madsen, H., Mahjour, J., & Gryseels, B. (2000a). New challenges in schistosomiasis control in Morocco. *Acta Tropica*, *77*(1), 61–67. doi:10.1016/S0001-706X(00)00114-5 PMID:10996121

Laamrani, H., & Lhayati, M. (1996). *Analyse préliminaire de la situation de la schistosomiase au foyerde Tata-Akka*. Rabat: Ministere de Sante Publique.

Laamrani, H., Mahjour, J., Madsen, H., Khallaayoune, K., & Gryseels, B. (2000b). Schistosoma haematobium in Morocco: Moving from control to elimination. *Parasitology Today (Personal Ed.)*, *16*(6), 257–260. doi:10.1016/S0169-4758(00)01665-3 PMID:10827435

Laaziri, M.(2012).Élimination de la schistosomiase au Maroc: Une réalité et un succès après trois décennies de lutte, OMS Bureau régional de la Méditerranée Orientale.

Laaziri, M., & Benouna, M. (1982). *Guide de la lutte contre la bilharziose. Direction des affaires techniques*. Rabat, Maroc: Ministe`re de la Sante Publique.

Lal, R. (2003). Soil erosion and the global carbon budget. *Environment International*, *29*(4), 437–450. doi:10.1016/S0160-4120(02)00192-7 PMID:12705941

Lebel, L., Rattanawilailak, S., & Lebel, P. (2010). Lazy Gardens – A Sustainability Experiment in the Uplands of Northern Thailand, Draft Working Paper 2, Unit for Social and Environmental Research. Faculty of Sciences, Chiang Mai University, Thailand.

Lee, M. G., & Terry, S. I. (1989). Arteriomesenteric duodenal occlusion associated with strongyloidiasis. *The Journal of Tropical Medicine and Hygiene*, *92*, 41–45. PMID:2918578

Lillesand, T., Kiefer, R., & Chipman, J. (2004). *Remote Sensing and Image Interpretation* (5th ed.). New York: John Wiley.

Lima, J. P., & Delgado, P. G. (1961). Diagnosis of strongyloidiasis: importance of Baermann's method. *American Journal of Digestive Diseases 6 (New Series):* 899-904.

Lindo, J. F., Conway, D. J., Atkins, N. S., Bianco, A. E., Robinson, R. D., & Bundy, D. A. P. (1994). Prospective evaluation of enzyme-linked immunosorbent assay and immunoblot methods for the diagnosis of endemic *Strongyloides stercoralis* infection. *The American Journal of Tropical Medicine and Hygiene*, *51*(2), 175–179. doi:10.4269/ajtmh.1994.51.175 PMID:8074251

Little, E., Langston, D., & Eaker, T. (2019). *Disease Management In The Home Vegetable Garden*, UGA Cooperative Extension Circular 862, Accessed on July 15, 2019.

Little, M. D. (1966). Comparative morphology of six species of Strongyloides (Nematoda) and redefinition of the genus. *The Journal of Parasitology*, *52*(1), 69–84. doi:10.2307/3276396 PMID:5929983

Liu, L. X., & Weller, P. F. (1993). Strongyloidiasis and other intestinal nematode infections. *Infectious Disease Clinics of North America*, *7*, 655–682. PMID:8254165

Lo, C. Y., & Kajioka, R. (1989). Predictive values. *American Journal of Clinical Pathology*, *91*(4), 505–506. doi:10.1093/ajcp/91.4.505b PMID:2597242

Longworth, D. L., & Weller, P. F. (1986). Hyperinfection syndrome with strongyloidiasis. In J. S. Remington, & M. N. Swartz (Eds.), *Current clinical topics in infectious diseases* (Vol. 7, pp. 1–26). New York: McGraw-Hill.

Machicado, J. D., Marcos, L. A., Tello, R., Canales, M., Terashima, A., & Gotuzzo, E. (2012). Diagnosis of soil-transmitted helminthiasis in an Amazonic community of Peru using multiple diagnostic techniques. *Transactions of the Royal Society of Tropical Medicine and Hygiene*, *106*(6), 333–339. doi:10.1016/j.trstmh.2012.03.004 PMID:22515992

Mahjour, J., Akalay, O., & Saddiki, A. (1992). *Les leishmanioses au Maroc de l'analyse écoépidémiologie à la prévention*. DEPS Bul Epidém.

Maire, N., Smith, T., Ross, A., Owusu-Agyei, S., Dietz, K., & Molineaux, L. (2006). A model for natural immunity to asexual blood stages of Plasmodium falciparum malaria in endemic areas. *The American Journal of Tropical Medicine and Hygiene*, *75*(2), 19–31. doi:10.4269/ajtmh.2006.75.19 PMID:16931812

Marambe, B., Weerahewa, H. L. J., Pushpakumara, D. K. N. G., Silva, G. L. L. P., Punyawardena, B. V. R., Premalal, S., Miah, G., & Roy, J. (2012). Vulnerability of Home garden Systems to Climate Change and its Impacts on Food Security in South Asia, Project Report. *Submitted to APN, (Project No. ARCP2010-03CMY-Marambe)*. Retrieved from www.apn-gcr.org/.../files/3ddf57b875774091a38f95bdfee0b6c9.pdf, accessed on September 10, 2012.

Marty, F. M., Lowry, C. M., Rodriguez, M., Milner, D. A., Pieciak, W. S., Sinha, A., ... Baden, L. R. (2005). Treatment of human disseminated strongyloidiasis with a parenteral veterinary formulation of ivermectin. *Clinical Infectious Diseases*, *41*(1). doi:10.1086/430827 PMID:15937753

Mattsson, E., Ostwald, M., Nissanka, S. P., & Marambe, B. (2013). Homegardens as a multi-functional land-use strategy in Sri Lanka with focus on carbon sequestration. *Ambio*, *42*(7), 892–902. doi:10.100713280-013-0390-x PMID:23456780

Mbuya, M. N. N., & Humphrey, J. H. (2016). Preventing environmental enteric dysfunction through improved water, sanitation and hygiene: an opportunity for stunting reduction in developing countries. *Maternal & Child Nutrition, 12*(Suppl Suppl 1), 106–120.

Medina, L. S., Heiken, J. P., & Gold, R. P. (1992). Pipestem appearance of small bowel in strongyloidiasis is not pathognomonic of fibrosis and irreversibility. *American Journal of Roentgenology, 159*(3), 543–544. doi:10.2214/ajr.159.3.1503020 PMID:1503020

Mejia, R., & Nutman, T. B. (2012). Screening, prevention, and treatment for hyperinfection syndrome and disseminated infections caused by *Strongyloides stercoralis. Current Opinion in Infectious Diseases, 25*(4), 458–463. doi:10.1097/QCO.0b013e3283551dbd PMID:22691685

Memon, N (2012). Malevolent Floods of Pakistan 2010-2012.

Messias, I. J. T., Genta, R. M., & Mohren, W. D. (1994). Adherence of monocytes and polymorphonuclear cells to infective larvae of *Strongyloides stercoralis* after complement activation. *The Journal of Parasitology, 80*(2), 267–274. doi:10.2307/3283757 PMID:8158470

Mike, S. (January 31, 2006). Beat neglected diseases to fight HIV, TB, and malaria. *SciDev.Net.* Archived from the original on May 19, 2006.

Mockenhaupt, F. P., Ehrhardt, S., Gellert, S., Otchwemah, R. N., Dietz, E., Anemana, S. D., & Bienzle, U. (2004). α+-thalassemia protects African children from severe malaria. *Blood, 104*(7), 2003–2006. doi:10.1182/blood-2003-11-4090 PMID:15198952

Molinari, G., Soloneski, S., & Larramendy, M. L. (2010). New ventures in the genotoxic and cytotoxic effects of macrocyclic lactones, abamectin and ivermectin. *Cytogenetic and Genome Research, 128*(1-3), 37–45. doi:10.1159/000293923 PMID:20389039

Moorthy, C. G., & Raj, S. I. (2018). Fixed points of sequences of mappings, *Results in Fixed Point Theory and Applications, 2018*(1), 1-10.

Morgan, J. (2005). Global trends in candidemia: A review of reports from 1995–2005. *Current Infectious Disease Reports, 7*(6), 429–439. doi:10.100711908-005-0044-7 PMID:16225780

MS . (2019). Ministère de la santé Maroc. Récupéré sur https://www.sante.gov.ma/Reglementation/Pages/Accueil.aspx]. *MD Medical Newsmagazine, 11*, 15.

Muennig, P., Pallin, D., Challah, C., & Khan, K. (2004). The cost-effectiveness of ivermectin vs. albendazole in the presumptive treatment of strongyloidiasis in immigrants to the United States. *Epidemiology and Infection, 132*(6), 1055–1063. doi:10.1017/S0950268804003000 PMID:15635962

Muriuki, J. M., Mentzer, A., Band, G., Gilchrist, J., Carstensen, T., Lule, S., & Cutland, C. (2019). The ferroportin Q248H mutation protects from anemia, but not malaria or bacteremia. *Science Advances, 5*(9), 1–12. doi:10.1126ciadv.aaw0109 PMID:31517041

Musa, S. S., Zhao, S., Chan, H. S., Jin, Z., & He, D. (2019). A mathematical model to study the 2014–2015 large-scale dengue epidemics in Kaohsiung and Tainan cities in Taiwan, China. *Mathematical Biosciences and Engineering*, *16*(5), 3841–3863. doi:10.3934/mbe.2019190 PMID:31499639

Musbah, E., & Worku, A. (2016). Influence of maternal education on child stunting in SNNPR, Ethiopia. *Central African Journal of Public Health*, *2*(2), 71.

NDMA. (2018). *Report on Prevailing Drought Like Situation in Sindh With*. (November). Retrieved from https://reliefweb.int/sites/reliefweb.int/files/resources/Drought Situation Report of Sindh Particularly in District Tharparkar-2018%2812-11-2018%29.pdf

Newfield, T. P. (2016). Mysterious and Mortiferous Clouds: The Climate Cooling and Disease Burden of Late Antiquity. *Late Antique Archaeology*, *12*(1), 89–115. doi:10.1163/22134522-12340068

Nguyen, T. (1997). A Study on the Home Garden Ecosystem in the Mekong River Delta and the Hochiminh City, Working Paper No. 7, UNESCO, Division of Ecological Sciences, South South Co-operation Programme. Retrieved from http://www.unesco.org/mab/activity/s-s/s-home.html

Nielsen, P. B., & Mojon, M. (1987). Improved diagnosis of *Strongyloides stercoralis* by seven consecutive stool specimens. Dntralblatt fur Bakteriologie. *Mikrobiologie und Hygiene*, *263*, 616–618. PMID:3604502

Nishtar S. (2007). Health Indicators of Pakistan – Gateway Paper II. Islamabad, Pakistan: Heartfile; 2007.UNICEF Pakistan Situation Report No. 21, Final Report on 2012 Monsoon Floods in Pakistan, April 7, 2013, and Humanitarian Action for Children (HAC) 2014 Report.

Nutman, T. B. (2006). Asymptomatic peripheral blood eosinophilia redux: Common parasitic infections presenting frequently in refugees and immigrants [editorial]. *Clinical Infectious Diseases*, *42*(3), 368–369. doi:10.1086/499247 PMID:16392082

Nutman, T. B., & Krolewiecki, A. (2019). Strongyloidiasis: A Neglected Tropical Disease. *Infectious Disease Clinics*, *33*(1), 135–151. PMID:30712758

OCHA. (2019). *Asia and the Pacific: Weekly Regional Humanitarian Snapshot (September 25 - October 1, 2018) No Title*. Retrieved from https://reliefweb.int/sites/reliefweb.int/files/resources/ROAP_Snapshot_190121.pdf

Ohnishi, K., Kogure, H., Kaneko, S., Kato, Y., & Akao, N. (2004). Strongyloidiasis in a patient with acquired immunodeficiency syndrome. *Journal of Infection and Chemotherapy*, *10*(3), 178–180. doi:10.100710156-004-0312-8 PMID:15290458

Olsen, A., van-Lieshout, L., Marti, H., Polderman, T., Polman, K., Steinmann, P., ... Magnussen, P. (2009). Strongyloidiasis–the most neglected of the neglected tropical diseases? *Transactions of the Royal Society of Tropical Medicine and Hygiene*, *103*(10), 967–972. doi:10.1016/j.trstmh.2009.02.013 PMID:19328508

Ordonez, L. E., & Angulo, E. S. (2004). Efficacy of ivermectin in the treatment of children parasitized by Strongyloides stercoralis. *Biomedica: revista del Instituto Nacional de Salud 24*(1): 33-41.

Ottesen, E. A., & Campbell, W. C. (1994). Ivermectin in human medicine. *The Journal of Antimicrobial Chemotherapy*, *34*(2), 195–203. doi:10.1093/jac/34.2.195 PMID:7814280

Oudada, M. (2009). L'eau dans le Bani: Disparités et gestion participative. In *Eaux, pauvreté et crises sociales*. Marseille: IRD Éditions; doi:10.4000/books.irdeditions.4893

Pakistan Health Cluster - Floods in Pakistan. (2010). (*No. 19, Focus on donors, September 28, 2010*). Cairo: World Health Organization Regional Office for the Eastern Mediterranean.

Palecki, M. A., Changnon, S. A., & Kunkel, K. E. (2001). The nature and impacts of the July 1999 heat wave in the Midwestern United States: Learning from the lessons of 1995. *Bulletin of the American Meteorological Society*, *82*(7), 1353–1367. doi:10.1175/1520-0477(2001)082<1353:TNAIOT>2.3.CO;2

Pezza, A. B., van Rensch, P., & Cai, W. (2012). Severe heat waves in Southern Australia: Synoptic climatology and large-scale connections. *Climate Dynamics*, *38*(1–2), 209–224. doi:10.100700382-011-1016-2

Politis, D. S., Triantafyllidi, E., Katsanos, K. H., & Christodoulou, D. K. (2017). Strongyloides stercoralis infection: An "exotic" cause of chronic diarrhea. *Annals of Gastroenterology*, *30*(4), 467. PMID:28655989

Prüss, A., Kay, D., Fewtrell, L., & Bartram, J. (2002). Estimating the burden of disease from water, sanitation, and hygiene at a global level. *Environmental Health Perspectives*, *110*(5), 537–542. PMID:12003760

Pushpakumara, G., Marambe, B., Silva, G. L. L. P., Weerahewa, J., & Punyawardena, R. (2012). Homegardens in Sri Lanka: The status importance and future perspective. *Tropical Agriculturist*, *159*, 55–125.

Requena-Mendez, A., Buonfrate, D., Bisoffi, Z., & Gutiérrez, J. M. (2014). Advances in the diagnosis of human strongyloidiasis. *Current Tropical Medicine Reports*, *1*(4), 440–475. doi:10.100740475-014-0034-7

Requena-Mendez, A., Chiodini, P., Bisoffi, Z., Buonfrate, D., Gotuzzo, E., & Muñoz, J. (2013). The laboratory diagnosis and follow up of strongyloidiasis: A systematic review. *PLoS Neglected Tropical Diseases*, *7*(1). doi:10.1371/journal.pntd.0002002 PMID:23350004

Research Publications. (July 23, 2017). Poverty Solutions at The University of Michigan. Retrieved from www.npc.umich.edu

Rguioui, I. (2006). Epidémiologie de la leishmaniose cutanée à SEFROU. Rabat: Thèse de doctorat en pharmacieFaculté de médecine et de pharmacie, Université Mohammed V-Rabat, Maroc.

Rioux, J. (2007). Le paradigme « écopathologie » Son application à l'épidémiologie des leishmanioses. Académie des sciences et lettres de Montpelier.

Rock, K. S., Ndeffo-Mbah, M. L., Castaño, S., Palmer, C., Pandey, A., Atkins, K. E., & Chitnis, N. (2018). Assessing strategies against Gambiense sleeping sickness through mathematical modeling. *Clinical Infectious Diseases*, *66*(4suppl_4), 286–292. doi:10.1093/cid/ciy018 PMID:29860287

Roger, D., & Gonzalez, E. M. N. (2017). Evaluacio'n del conocimiento sobre enfermdades de sexto curso del grado en Medicina de la Comunidad de Madrid. *IV Congreso de estudiantes de Medicina Universidad Rey Juan Carlos; May 25, 2017*. Madrid, Spain: Universidad Rey Juan Carlos.

Rotman, H. L., Yutanawiboonchai, W., Brigandi, R. A., Leon, O., Nolan, T. J., Schad, G. A., & Abraham, D. (1994). *Strongyloides stercoralis*: eosinophil dependent immune-mediated killing of infective L3 in Balb-kBYJ mice. *American Journal of Tropical Medicine and Hygiene. Abstracts:* 234.

Rozendaal, J. A. (1997). *Vector control: methods for use by individuals and communities*. World Health Organization.

Ruiz, M. J. A., Pons, M. J. A., Bermejo, L., Opez, J., Felices, A. F., Serra, S. J. A., & Felices, A. J. M. (1987). Estrongiloidiasis diseminada: Estudio clinico-patologico de un caso diagnosticado en autopsia. *Revista Clínica Española*, *180*, 381–384. PMID:3602552

Sadqi, M. (2013). Profil épidémiologique des leishmanioses cutanées dans la province de Ouarzazate. Thèse de doctorat en pharmacie, Faculté de médecine et de pharmacie, Université Mohammed V-Souissi, Maroc.

Safdar, A., Malathum, K., Rodriguez, S. J., Husni, R., & Rolston, K. V. I. (2004). Strongyloidiasis in patients at a comprehensive cancer centre in the United States: A retrospective study covering the years 1971–2003. *Cancer*, *100*, 1531–1536. doi:10.1002/cncr.20120 PMID:15042689

Saikia, P., Choudhury, B. I., & Khan, M. L. (2012). Floristic composition and plant utilization pattern in home gardens of Upper Assam, India. *Tropical Ecology*, *53*(1), 105–118.

Sandground, J. H. (1926). Biological studies on the life-cycle in the genus *Strongyloides grassi*, 1879. *American Journal of Hygiene*, *6*, 337–388.

Sandosham, A. A. (1952). An investigation into the association of creeping eruption with Strongyloides infection contracted in the Far East. *Journal of Helminthology*, *26*(1), 1–24. doi:10.1017/S0022149X00032569

Satoh, M., & Kokaze, A. (2004). Treatment strategies in controlling strongyloidiasis. *Expert Opinion on Pharmacotherapy*, *5*(2), 293–301. doi:10.1517/14656566.5.11,2293 PMID:15500376

Sato, Y., Inoue, F., Matsuyama, R., & Shiroma, Y. (1990). Immunoblot analysis of antibodies in human strongyloidiasis. *Transactions of the Royal Society of Tropical Medicine and Hygiene*, *84*(3), 403–406. doi:10.1016/0035-9203(90)90337-E PMID:2260175

Sato, Y., & Shiroma, Y. (1989). Peripheral lymphocyte subsets and their responsiveness in human strongyloidiasis. *Clinical Immunology and Immunopathology*, *53*(3), 430–438. doi:10.1016/0090-1229(89)90005-6 PMID:2572368

Scaglia, M., Brustia, R., Gatti, S., Bernuzzi, A. M., Strosselli, M., Malfitano, A., & Capelli, D. (1984). Autochthonous strongyloidiasis in Italy: An epidemiological and clinical review of 150 Cases. *Bulletin de la Société de Pathologie Exotique*, *77*, 328–332. PMID:6488423

Schaefer, M. (1993). *Health, environment, and development. Approaches to drafting country-level strategies for human well-being under Agenda 21 (No. WHO/EHE--93.1)*. World Health Organization.

Schaffer, A., Muscatello, D., Broome, R., Corbett, S., & Smith, W. (2012). Emergency department visits, ambulance calls, and mortality associated with an exceptional heat wave in Sydney, Australia, 2011: A time-series analysis. *Environmental Health*, *11*(1), 3. doi:10.1186/1476-069X-11-3 PMID:22273155

Scowden, E. B., Schaffner, W., & Stone, W. J. (1978). Overwhelming strongyloidiasis. An unappreciated opportunistic infection. *Medicine; Analytical Reviews of General Medicine, Neurology, Psychiatry, Dermatology, and Pediatries*, *57*(6), 527–544. doi:10.1097/00005792-197811000-00004 PMID:362122

Sen Roy, S., & Balling, R. (2004). Trends in Extreme Daily Precipitation Indices in India. *International Journal of Climatology*, *24*(4), 457–466. doi:10.1002/joc.995

Serfling, R. E. (1963). Methods for current statistical analysis of excess pneumonia-influenza deaths. *Public Health Reports*, *78*(6), 494–506. doi:10.2307/4591848 PMID:19316455

Shirley, M. D., & Rushton, S. P. (2005). The impacts of network topology on disease spread. *Ecological Complexity*, *2*(3), 287–299. doi:10.1016/j.ecocom.2005.04.005

Siddiqui, A. A., & Berk, S. L. (2001). Diagnosis of *Strongyloides stercoralis* infection. *Clinical Infectious Diseases*, *33*(7), 1040–1047. doi:10.1086/322707 PMID:11528578

Smith, T., & Weeks, E. (2019). Disease Management In The Home Vegetable Garden Agriculture And Landscape Program 4/09, Accessed on July 15,2019.

Solberg, K. (2010, October). Solberg (2010). Worst floods in living memory leave Pakistan in paralysis. *Lancet*, *376*(9746), 1039–1040. doi:10.1016/S0140-6736(10)61469-9 PMID:20879089

Spinicci, M., Macchioni, F., Mantella, A., Gabrielli, S., Roselli, M., Rojo-Mayaregua, D., ... Bartoloni, A. (2017). Seroepidemiological trend of strongyloidiasis in the Bolivian Chaco (1987–2013) in the absence of disease-specific control measures. *Tropical Medicine & International Health*, *22*(11), 1457–1462. doi:10.1111/tmi.12977 PMID:28929562

Sprott, V., Selby, C. D., Ispahani, P., & Toghill, P. J. (1987). Indigenous strongyloidiasis in Nottingham. *British Medical Journal (Clinical Research Ed.)*, *294*(6574), 741–742. doi:10.1136/bmj.294.6574.741-a PMID:3105720

Steer, L., Goudet, M., Akchour, M., & Mouradi Abdellah, H. (2004). *Les plantes aromatiques, médicinales et tinctoriales, un atout pour le développement rural de la région de Tata. Centre National d'Etudes Agronomiques des Régions Chaudes. En collaboration avec Direction Provinciale de l'Agriculture de Tata et ALCESDAM (Maroc). Edition: P. JOUVE, C.* France: SEUGE.

Suputtamongkol, Y., Premasathian, N., Bhumimuang, K., Waywa, D., Nilganuwong, S., Karuphong, E., ... Silpasakorn, S. (2011). Efficacy and safety of single and double doses of ivermectin versus 7-day high dose albendazole for chronic strongyloidiasis. *PLoS Neglected Tropical Diseases*, *5*(5). doi:10.1371/journal.pntd.0001044 PMID:21572981

Tangjang, S., & Arunachalam, A. (2009). Role of traditional home garden systems in North East India, *Indian Journal of Traditional Knowledge, January,* *8*(1), pp. 47–50.

Thompson, J. R., & Berger, R. (1991). Fatal adult respiratory distress syndrome following successful treatment of pulmonary strongyloidiasis. *Chest*, *99*(3), 772–774. doi:10.1378/chest.99.3.772 PMID:1995245

Tjaden, N. B., Caminade, C., Beierkuhnlein, C., & Thomas, S. M. (2018). Mosquito-borne diseases: Advances in modelling climate-change impacts. *Trends in Parasitology*, *34*(3), 227–245. doi:10.1016/j.pt.2017.11.006 PMID:29229233

Uejio, C. K., Wilhelmi, O. V., Golden, J. S., Mills, D. M., Gulino, S. P., & Samenow, J. P. (2011). Intra-urban societal vulnerability to extreme heat: The role of heat exposure and the built environment, socioeconomics, and neighborhood stability. *Health & Place*, *17*(2), 498–507. doi:10.1016/j.healthplace.2010.12.005 PMID:21216652

Vazquez-Guillamet, L. J., Saul, Z., Miljkovich, G., Vilchez, G. A., Mendonca, N., Gourineni, V., ... Gangcuangco, L. M. (2017). *Strongyloides stercoralis* infection among human immunodeficiency virus (HIV)-infected patients in the United States of America: A case report and review of literature. *The American Journal of Case Reports*, *18*, 339–346. doi:10.12659/AJCR.902626 PMID:28366929

Viera, R. A. (1979). *"Strongyloides stercoralis."* Um nematodo de relevante importancia em medicina humana. *Jornal do Medico C*, *1838*, 109–126.

Warraich, H., Zaidi, A. K. M., & Patel, K. (2011). Floods in Pakistan: A public health crisis. *Bulletin of the World Health Organization*, *89*(3), 236–237. doi:10.2471/BLT.10.083386 PMID:21379421

Warwick, R. A. (1975). Drought hazard in the United States: A research assessment: Boulder, Colorado, University of Colorado, Institute of Behavioral Science, Monograph no. NSF/RA/E-75/004.

Watson, J. M. (1958). Ecology and distribution of Bulinus truncatus in the Middle East; with comments on the effect of some human activities in their relationship to the snail host on the incidence of bilharziasis haematobia in the Middle East and Africa. *Bulletin of the World Health Organization*, *18*(5-6), 833–894. PMID:13573115

Weerahewa, J., Pushpakumara, G., Silva, P., Daulagala, C., Punyawardena, R., Premalal, S., ... & Marambe, B. (2012). Are homegarden ecosystems resilient to climate change? An analysis of the adaptation strategies of homegardeners in Sri Lanka. *APN Science Bulletin*, *2*(2), 22–27.

WHO and UNICEF. (1978). Declaration of Alma-Ata. *WHO Chronicle*, *32*, 428–430. PMID:11643481

WHO. (1957). Study Group on the Ecology of Intermediate Snail Hosts of Bilharziasis, World Health Organization, Study Group on the Ecology of Intermediate Snail Hosts of Bilharziasis [meeting held in Paris from October 3 to 9, 1956] : report. World Health Organization technical report series; No. 120.

WHO. (2017). *Preventive chemotherapy to control soil-transmitted helminth infections in at-risk population groups* (p. 87). Geneva, Switzerland: WHO.

WHO. (2019, Dec. 25). World Health Organization. Récupéré sur World Health Organization: https://www.who.int/leishmaniasis/surveillance/en/

Wilhite, D. A., & Lantz, M. H. (1985). Understanding the drought phenomenon: The role of definitions. *Water International*, *103*(3), 111–120. doi:10.1080/02508068508686328

WMO (World Meteorological Organization). (1975). *Drought and Agriculture. Report of the CAgM Working Group on the Assessment of Drought prepared by C. E. Hounam Chairman, J. J. Burgos, M. S. Kalik, W. C. Palmer, & J. Rodda. Technical Note No. 138, WMO No. 392*. Geneva, Switzerland: World Meteorological Organization.

Woodring, J. H., Halfhill, H. I. II, & Reed, J. C. (1994). Pulmonary strongyloidiasis: Clinical and imaging features. *AJR. American Journal of Roentgenology*, *162*(3), 537–542. doi:10.2214/ajr.162.3.8109492 PMID:8109492

Xu, Y., Dadvand, P., Barrera-Gómez, J., Sartini, C., Marí-Dell'Olmo, M., Borrell, C., ... Basagaña, X. (2013). Differences on the effect of heat waves on mortality by sociodemographic and urban landscape characteristics. *Journal of Epidemiology and Community Health*, *67*(6), 519–525. doi:10.1136/jech-2012-201899 PMID:23443960

Yacoubi, B., Zekhnini, A., Rondelaud, D., Vignoles, P., Dreyfuss, G., Cabaret, J., & Moukrim, A. (2007). Habitats of Bulinus truncatus and Planorbarius metidjensis, the intermediate hosts of urinary schistosomosis, under a semiarid or an arid climate. *Parasitology Research*, *101*(2), 311–316. doi:10.100700436-007-0500-4 PMID:17340142

Zaha, O., Hirata, T., Kinjo, F., & Saito, A. (2000). Strongyloidiasis--progress in diagnosis and treatment. *Internal Medicine (Tokyo, Japan)*, *39*(9), 695–700. doi:10.2169/internalmedicine.39.695 PMID:10969899

Zaitchik, B. F., Macalady, A. K., Bonneau, L. R., & Smith, R. B. (2006). Europe's 2003 heat wave: A satellite view of impacts and land–atmosphere feedbacks. *International Journal of Climatology*, *26*(6), 743–769. doi:10.1002/joc.1280

About the Contributors

Ahmed Karmaoui started his academic career as Research Fellow at Cadi Ayyad University and awarded Ph.D. from Faculty of Semlalia; and as an associate editor at the International Journal of Climate Change: Impacts and Responses, editor at several International Journals such as Journal of Research in Chemistry & Environment (SRI: JRCE), the International Journal of Earth & Environmental Sciences, Nanomedicine and Nanoscience Research Journal; as International Advisory Board member of the International Journal of ICT Research and Development in Africa (IJICTRDA), and the International Journal of Social Ecology and Sustainable Development (IJSESD). He is a Member of the Asian Council of Science Editors. KARMAOUI has reviewed for 63 international journals, graduated from Publons Academy, and awarded Top 1% of international Reviewers (2018) by Publons (part of CLARIVATE, supported by WEB OF SCIENCE). He has published various peer-reviewed international journal articles and book chapters on a range of environmental topics. He is currently the president of the Southern Center for Culture and Science, Morocco. Broad areas of his research interests include ecosystem services assessment.

Abdelkrim Ben Salem is an Assistant Professor of Ecology and Environment in the Mohammed V University, Faculty of Sciences Rabat, Department of Biology. He joined the University of Mohammed V after ten years as a teacher of qualifying secondary education in Draa Tafilalet Academy. Broadly, her methodological research focuses on vulnerability and adaptation to climate change and valuation of ecosystem services. Within, Ben Salem is currently working on socioeconomic vulnerability of the agriculture and water sector to climate change. He has published interdisciplinary projects across varied outlets, including, Parasites & Vectors Journal, International Journal of Water Resources and Arid Environments and International Publisher of Information Science and Technology Research (IGI-Global).

Ashfaq Ahmad Shah is an Assistant Professor of Economics at the School of Social Sciences and Humanities.

* * *

Ali Ait Boughrous is a Professor of Hydrobiology and Environmental sciences in the Moulay Ismail University, Faculty of Sciences and Techniques, Errachidia, Morocco.

Muhammad Ali is Assistant Professor, Department of Economics, School of Social Sciences and Humanities, National University of Sciences and Technology, Islamabad, Pakistan.

Souad Ben Salem has a master's degree in Sciences and Technics at the Faculty of Sciences and Technics in Errachidia Morocco, and is a student at the Faculty of Sciences Semlalia Marrakech, Doctoral Training: Science of Life and Environment, laboratory of hydrobiology, ecotoxicology, sanitation and global changes.

Sarin Ishaque received her BS degree in Economics from National University of Science and Technology in 2019. Same year, she was hired as the Assistant Product Manager in DKT International, a NGO working in Pakistan since 2015. It is one of the biggest family planning services provider globally as of now. She is currently residing in Karachi and working in their head office. Before this, she interned in British American Tobacco and Unilever and worked in various customer development projects and gained a good market insight. She is trying to create an impact, as little, as big where necessary and is on a quest to improve the lives of women and children for a better tomorrow.

Sebak Kumar Jana is currently Professor of Economics in the Department of Economics, Vidyasagar University, West Bengal, India. Dr. Jana was graduated from Presidency College, Kolkata in 1989 and did his post graduation in Economics from University of Calcutta in 1991. He obtained M.Phil. and Ph.D. degree in Economics from Jadavpur University, Kolkata. His major area of academic interest includes environmental and resource economics, rural development and economics of education. He has about eighty publications in Journals and in edited books. He has published six books as author or editor including Tank Irrigation in the Dry Zones in India, Education in West Bengal – Looking Beyond Schools, Development in Development in Developing Economies (Edited). He, as Principal Investigator, has completed a few research projects funded by ICSSR, UGC and NABARD and IWMI.

Junaid Ahmad Malik has completed his PhD in Zoology (Wildlife) in 2015. He is working as a Lecturer in Zoology at Govt. Degree College, Bijbehara, Anantnag. Dr. Malik has published many research articles and technical papers in many International and National reputed Journals. He has also authored several books and book chapters of reputed publishers. He is in the editorial board and a regular reviewer of several journals of reasonable reputation.

C. Ganesa Moothy is being a professor in mathematics. He has more than 34 years of teaching experience. He has taught Engineering students, Science students and Mathematics students. He has taught all applied mathematics subjects like differential equations, numerical methods, statistics, and he has taught all pure mathematics subjects like algebra, analysis, number theory. He has published more than 72 research articles, and 4 books. Fourteen students have completed their Ph.D. degree under his guidance. He has solved a 50-year-old open problem in an article published in the journal "Mathematika" in 1992. He is interested at present in Basic physics, Basic Mathematics and Biological differential equations modeling.

Adil Moumane is a Researcher in the department of Geography. Lab. Environment, Societies, and Territories. Ibn Tofail University, Kenitra, Morocco.

G. Udhaya Sankar is a researcher in theoretical physics and applied physics. He is also being a subject expert and published articles in climate changes. He has published articles regarding environmental sciences and pollution control. He is an author of nineteen research articles. He is being a reviewer in the following reputed journals: Applied Energy; Fundamental Journal of Mathematics and Applications; and Recent innovations in chemical engineering. He is being member of Editorial Team in five international journals from various countries like Singapore, Malaysia, UAE, UK, and USA. He has written the following three books: Planets And Electromagnetic waves, First Edition, Idea Publishing, New Delhi, India. (2018) ISBN:9789386518835, 938651883X. Mathematics in Material Science, First Edition, Lambert Academic Publishing, Germany, 2019. ISBN: 9786200084248, 6200084246, Numerical Methods for Calculus Students, First Edition, Lambert Academic Publishing, Germany, (2019) ISBN: 9786139909865, 6139909864.

Junaid Ul Mulk received his bachelor's degree in BS Economics from National University of Science and Technology in 2019. In 2018 he was hired as the only Student Marketeer from NUST to join Redbull Pakistan. He is currently working in HBL as a Management Trainee Officer.

Index

www.ingramcontent.com/pod-product-compliance
Lightning Source LLC
LaVergne TN
LVHW061533290926
849162LV00004B/782